Silver Burdett Ginn Mathematics

Daily Review

Practice

Problem Solving

Mixed Review

4

Silver Burdett Ginn

Parsippany, NJ

Atlanta, GA • Deerfield, IL • Irving, TX • Needham, MA • Upland, CA

Silver Burdett Ginn
299 Jefferson Road, P.O. Box 480
Parsippany, NJ 07054-0480

1999 Printing

ISBN 0-382-37319-7

15 16 17 18 19 -PO- 05 04 03 02

Contents

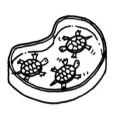

Chapter 1 1

Chapter 2 11

Chapter 3 24

Chapter 4 39

Chapter 5 51

Chapter 6 61

Chapter 7 70

Chapter 8 85

Chapter 9 97

Chapter 10 111

Chapter 11 122

Chapter 12 136

EXPLORE: Understanding Place Value

Complete the chart to show the value of each number.

The first one is done for you.

Place	Thousands	Hundreds	Tens	Ones
1. Number	7	7	7	7
Value	7,000 or 7 thousands			
2. Number	4	5	1	1
Value				
3. Number	9	8	9	7
Value				

Problem Solving

Solve.

4. If Abbey has $222.22 in her savings account, does she have more than or less than $222.33? _____

5. How much more than or less than $222.33 does Abbey have? _____

6. If Abbey earns $1.17 in interest, then how much money does she have in her savings account? _____

Review and Remember

Add or subtract.

7. 12 + 14 = _____

8. 15 − 2 = _____

9. 28 + 2 = _____

10. 17 − 5 = _____

11. 4 + 15 = _____

12. 49 − 3 = _____

Name _____

Place Value to Hundred Thousands

Write each number in standard form.

1. six thousand, seven hundred five _____

2. fifty-five thousand _____

3. 700,000 + 60,000 + 4,000, + 900 + 23 _____

Write each number in expanded form. Then write the value of the underlined digit.

4. 3<u>5</u>,837 _____

5. 52<u>6</u>,790 _____

6. <u>4</u>93,805 _____

Problem Solving

Use the table to solve these problems.

Score Card	
Player	**Prize Points**
René	700,000
Shateel	26,000
Jennifer	263,000
Paul	6,000
Liv	500,000

7. Which two players have a total of 289,000 points?

_____

8. Which player has two hundred sixty-three thousand points?

_____

Review and Remember

Add or subtract.

9.
$$5$$
$$+\ 10$$

10.
$$4$$
$$8$$
$$+\ 2$$

11.
$$45$$
$$-\ 7$$

12.
$$63$$
$$-\ 9$$

Comparing and Ordering Numbers

Compare. Use < or >.

1. 4,374 ◯ 4,747   **2.** 536 ◯ 566   **3.** 811,569 ◯ 801,589

4. 12,463 ◯ 12,433   **5.** 1,355 ◯ 1,453   **6.** 14,597 ◯ 14,497

7. 103,601 ◯ 1,033,006 **8.** 9,019 ◯ 9,119   **9.** 119,898 ◯ 129,898

Order each set from greatest to least.

10. 558   323   342   _____

11. 11,632   11,649   11,610   _____

12. 213,080   213,830   231,820   _____

Problem Solving

Use the information in the table to help you solve these problems.

Favorite Toppings			
Location	Chocolate	Marshmallow	Butterscotch
Mall	7,350	7,315	3,700
Downtown	26,340	26,450	2,600
Beach	6,000	6,310	300
Park	53,400	5,340	534

13. Which topping is most popular at the Mall location?

_____

14. How many marshmallow toppings were sold at the Downtown, Beach, and

Park locations? _____

Review and Remember

Write the missing numbers in each pattern.

15. 15, 20, _____, 30   **16.** 14, _____, 20, 23   **17.** 30, 60, 90, _____

Use with Grade 4, text pages 8-11.   **3**

Problem Solving
Exact Numbers or Estimates

Answer each question. Give a reason for your choice.

Melissa and her family drove to Florida on vacation. The trip covered just 1,198 miles and they drove about four hundred miles each day. When they arrived, it was sunny with a temperature of 87°F.

1. Which of these numbers is an estimate? Explain your choice.

    a. 1,198 miles

    b. about 400 miles

    c. 87°F

_____

2. Which of the following is true?

    a. Melissa's family drove the same distance each day.

    b. The trip took three days.

    c. They made the trip in a van.

3. If Melissa's family left home on Saturday, what day did they arrive in Florida?

_____

Review and Remember

Compare the numbers. Choose $<$ or $>$.

4. 1,012 $\bigcirc$ 998   **5.** 5,290 $\bigcirc$ 5,902   **6.** 2,899 $\bigcirc$ 2,846

Round to the place of the underlined digit.

7. 8̲5   **8.** 7̲38   **9.** 3̲,435   **10.** 7,5̲82

_____   _____   _____   _____

EXPLORE: Millions

Answer these questions.

1. How many thousands are in one hundred thousand? _____

2. How many hundred thousands are in one million? _____

3. How many hundreds are in ten thousand? _____

4. How many ones are in ten thousand? _____

5. Why is 1,000,000 larger than 100,000?

_____

_____

Problem Solving

Use what you know about the number 1,000,000 to help you solve these problems.

6. Ten tens are 100. What are one hundred hundreds? _____

7. What are one thousand thousands? _____

Review and Remember

Add or subtract.

8. 12 + 13 = _____   **9.** 47 − 14 = _____   **10.** 29 + 13 = _____

11. 8 + 17 = _____   **12.** 26 − 3 = _____   **13.** 77 − 32 = _____

14. 15 − 5 = _____   **15.** 47 − 20 = _____   **16.** 13 + 9 = _____

Multiply or divide.

17. 7 × 9 = _____   **18.** 4 × 8 = _____   **19.** 64 ÷ 8 = _____

20. 72 ÷ 9 = _____   **21.** 36 ÷ 4 = _____   **22.** 19 × 0 = _____

23. 15 ÷ 1 = _____   **24.** 11 × 2 = _____   **25.** 56 ÷ 8 = _____

Place Value Through Millions

Write the standard form for each number.

1. four million, six hundred thousand _____

2. six hundred fifty-five million _____

Give the value of 4 in each number.

3. 2,435,837 _____   **4.** 224,759,600 _____

Problem Solving

Use the table to solve these problems.

Table 1: Servings of Butter Sold	
District	**Servings**
Shawsheen	5,475,700
Suffolk	7,890,900
King County	7,809,900
San Pedro	4,979,450

Table 2: Servings of Butter Sold	
District	**Servings**

5. Use Table 2 to order the amounts of servings from the least to the most.

6. Which districts sold between 5 million and 8 million servings of butter?

_____

Review and Remember

Use mental math or paper and pencil to subtract. Tell which method you use.

7.   15
        $-\,10$

8.   14
        $-\,8$

9.   79
        $-\,7$

10.   59
         $-\,9$

_____   _____   _____   _____

_____   _____   _____   _____

EXPLORE: Finding Patterns in Numbers

Look at the numbers in the box.

4,777	6,777	9,777	7,777

1. What is the same about these numbers?

_____

2. What is different about these numbers?

_____

3. List the numbers from least to greatest. Then fill in the numbers that are needed to complete the pattern.

_____

What are the next three numbers in this pattern?

4. 12,666   13,666   14,666   _____   _____   _____

Problem Solving

Use what you know about patterns to help you solve these problems.

5. Sally is making a necklace with beads. How should she continue this pattern:

30 red, 50 green, 70 red? _____

6. Miguel thinks that these numbers form a pattern:

6,666   666   66   6

Do you agree with Miguel? Tell why or why not.

_____

_____

Review and Remember

Find the patterns. Write the missing numbers.

7. 14, _____, 12, 11

8. 50, 70, _____, 110

9. 199, 299, 399, _____

10. 50, _____, 46, 44

Name _____

Problem Solving
Find a Pattern

Find a pattern. Use the pattern to solve each problem.

1. Tom has a total of 15 quarters and nickels. He places them in a row in the pattern quarter, nickel, quarter, nickel, and so on. The completed row has a quarter on each end. How many nickels does Tom have?

_____

2. Angela's garden is 30 feet long and 24 feet wide. If she fences in her garden using 6-foot sections of fence, how many fence posts will she need?

_____

3. Think about the addresses of the houses in your neighborhood. What patterns do you notice?

_____

4. Astronomers reported seeing Halley's Comet in the years 1833, 1909, and 1985. What pattern do you see here?

_____

Review and Remember

Write each number in word form.

5. 87   **6.** 953   **7.** 812   **8.** 1,606

Write the value of the underlined digit.

9. 52<u>3</u>   **10.** <u>9</u>12   **11.** 8<u>0</u>8   **12.** <u>7</u>,303

Rounding Numbers

Name the two tens, hundreds, or thousands that each number is between.

1. 78 _____

2. 689 _____

3. 7,901 _____

Round to the underlined place.

4. 8̲4 _____   **5.** 1̲27 _____   **6.** 78̲2 _____

7. 3̲45 _____   **8.** 1̲,829 _____   **9.** 7,0̲93 _____

10. 89̲3 _____   **11.** 9,69̲6 _____   **12.** 3,79̲9 _____

Problem Solving

Use the table to solve these problems. Round the numbers.

Winter Carnival	
Activity	**Number of People**
Sledding	65
Skating	235
Snow Sculpture	1,365
Total Attendance	1,665

13. Round to the nearest hundred: Total attendance at the Winter Carnival

was about _____ .

14. Round to the nearest ten: About _____ went skating.

Review and Remember

Write the value of the underlined digit.

15. 5,5̲5̲5 _____   **16.** 8̲,249 _____   **17.** 2,4̲07 _____

18. 61̲3 _____   **19.** 2,10̲5 _____   **20.** 1̲87 _____

Problem Solving
Using a Bar Graph

Use the bar graph below to answer the questions.

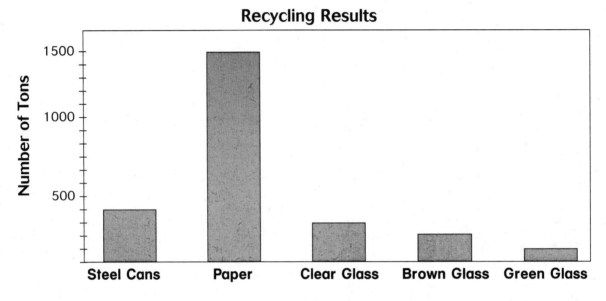

Recycling Results

1. How many tons of paper were collected? _____

2. How many tons of clear glass were collected? _____

3. How many tons of steel cans were collected? _____

4. Was more green glass or brown glass collected? _____

5. How many tons of materials were collected in all? _____

6. If the city wants to double its paper collection, how many tons will they have

to collect next year? _____

Review and Remember

Find the missing addend.

7. $5 + \underline{\hspace{1cm}} = 11$ **8.** $9 + \underline{\hspace{1cm}} = 16$ **9.** $\underline{\hspace{1cm}} + 8 = 14$

10. $3 + 3 + \underline{\hspace{1cm}} = 9$ **11.** $\underline{\hspace{1cm}} + 2 + 4 = 9$ **12.** $3 + \underline{\hspace{1cm}} + 5 = 8$

Addition and Subtraction Facts

Add or subtract.

1. 4 + 3	**2.** 18 − 9	**3.** 6 + 7	**4.** 16 − 8	**5.** 7 + 5
6. 5 + 9	**7.** 6 + 4	**8.** 11 − 2	**9.** 13 − 6	**10.** 9 + 3

Complete each fact family.

11. $8 + 9 = $ _____

$9 + 8 = $ _____

$17 − 9 = $ _____

$17 − 8 = $ _____

12. $8 + 6 = $ _____

$6 + 8 = $ _____

$14 − 6 = $ _____

$14 − 8 = $ _____

13. $7 + $ _____ $ = 10$

$3 + 7 = $ _____

$10 − $ _____ $ = 7$

_____ $ − 7 = 3$

14. _____ $ + 7 = 15$

$7 + $ _____ $ = 15$

_____ $ − 7 = 8$

$15 − 8 = $ _____

Problem Solving

15. Jay signed up to shoot baskets at Olympic Day. Each basket made counts for 1 point. If he scored 4 shots on the first round, and 9 shots on the second round, then what is his total score for both rounds? _____

16. Jennifer signed up to bounce a ball at Olympic Day. If she bounced the ball 7 times on the first round, and 6 times on the second round, then how many times did she bounce the ball altogether? _____

Review and Remember

Write the place value for the digit 7 in each number.

17. 749 _____

18. 17,890 _____

19. 7,344,680 _____

20. 475,846 _____

Estimating Sums and Differences

Estimate by rounding to the greatest place value.

1. 27
  + 43

2. 98
  − 19

3. $3.56
  + 7.22

4. 516
  − 182

Estimate by rounding to the nearest ten.

5. 53
  + 19 ___

6. 26
  + 14 ___

7. 44
  − 29 ___

8. 73
  − 36 ___

9. 72
  + 33 ___

Estimate by rounding to the nearest hundred.

10. 343
  + 219 ___

11. 264
  + 411 ___

12. 243
  − 219 ___

13. 743
  − 336 ___

Estimate by rounding to the nearest dollar.

14. $2.18
  +4.19 ___

15. $4.69
  − 2.82 ___

16. $4.27
  − 1.86 ___

Problem Solving

17. Tiffany wants to buy a necklace for $6.54 and a bracelet for $3.39.

Estimate what they will cost altogether. _____

18. Eduardo scored 467 points in a board game. His brother scored 359 points.

About how many points more did Eduardo score? _____

Review and Remember

Add or subtract.

19. $7 + 4 =$ ___

20. $16 - 8 =$ ___

21. $8 + 9 =$ ___

22. $14 - 7 =$ ___

Problem Solving

Is an Estimate Enough?

Silver Hill School is planning their field day for about 500 students in grades 3, 4, and 5. They have $2,458 to spend on food and entertainment. The field day will start on time at 9 o'clock and end at about 2 o'clock. The school expects 20–25 volunteers to help during the day.

Reread the paragraph above. Next to each item below, write whether it is an estimate or an exact amount. Explain.

1. Number of students _____

2. Number of grades _____

3. Money to spend _____

4. Starting time _____

5. Ending time _____

6. Number of volunteers _____

Review and Remember

Write each number in standard form.

7. five hundred thirty   **8.** seven thousand six   **9.** nine hundred five

_____   _____   _____

Order the numbers from least to greatest.

10. 317 409 356 289   **11.** 498 487 488 496   **12.** 3,480 3,469 2,453

_____   _____   _____

_____   _____   _____

Adding Two- and Three-Digit Numbers

Estimate first. Then add to find each sum.

1. 46
    + 96 _____

2. 78
    + 28 _____

3. 324
    + 758 _____

4. 639
    + 18 _____

Find each sum.

5. 253
    + 197

6. 686
    + 19

7. 44
     29
    + 73

8. 712
     133
    + 451

9. 342
    + 629

10. 534
    + 648

11. 243
    + 619

12. 743
     336
    + 712

Problem Solving

13. Gabriella earned $4.40 baby-sitting and $7.70 raking. How much did she

earn altogether? _____

14. Josh wants to plant a garden. He has 36 flower seeds and 72 vegetable
seeds. About how many more vegetable seeds than flower seeds does he

have? _____

Review and Remember

Compare. Write > or < .

15. 1,347 ◯ 1,224

16. 16,783 ◯ 17,899

17. 144,989 ◯ 411,999

18. 2,089 ◯ 2,890

19. 15,912 ◯ 15,921

20. 309,472 ◯ 390,247

Adding Greater Numbers

Estimate first. Then add to find each sum.

1. 3,446
      + 9,686 _____

2. 72,189
      + 28,556 _____

3. 2,324
        1,758 _____
      + 4,238

4. 6,432
      + 787 _____

5. 3,598
        4,866 _____
      + 4,977

6. 8,650
        2,228 _____
      + 998

Add. Use a calculator.

7. 89,473
      + 59,393

8. 93,865
        5,789
      + 645

9. 15,908
        45,897
      + 23,798

10. 56,895
         4,445
         821
       + 75,966

Problem Solving

Use the table to answer these problems.

Score Sheet	
Game	**Score**
Sean	35,908
Chris	22,970
Juan	14,988
Tamika	43,063
Yo	12,309

11. Juan and Chris are partners.

   What is their total score? _____

12. About how many more points does

   Tamika have than Yo? _____

Review and Remember

Write each number in standard form.

13. seven hundred forty-five thousand, six hundred five _____

14. nine hundred thousand, four hundred eight _____

Subtracting Two- and Three-Digit Numbers

Estimate first. Subtract. Add to check each answer.

1. 72 – 28 ____	**2.** 324 – 19 ____	**3.** 414 – 146 ____	**4.** 843 – 125 ____

Subtract.

5. 74 – 34	**6.** 85 – 3	**7.** 705 – 632	**8.** 827 – 369	**9.** $4.31 – 1.89
10. 932 – 486	**11.** 95 – 49	**12.** $8.63 – 5.29	**13.** 844 – 29	**14.** 894 – 347

Problem Solving

15. Sal read 876 pages in the library's Read-a-Thon. Anthony read 349 pages.

How many more pages did Sal read? _____

16. Kristen is buying a pound of oranges for $1.29 per pound. Terry is buying grapes for $2.19. Which costs more? How much more expensive is it?

_____

Review and Remember

Round to the nearest ten.

17. 47 _____   **18.** 784 _____

19. 894 _____   **20.** 5,870 _____

21. 509 _____   **22.** 6,372 _____

23. 4,988 _____   **24.** 7,032 _____

Subtracting Greater Numbers

Subtract.

1.   5,897
      − 4,977

2.   8,066
      − 4,782

3.   74,239
      − 14,855

4.   40,498
      − 13,899

5.   7,099
      − 5,887

6.   68,999
      − 53,099

7.   5,089
      − 4,866

8.   4,312
      − 1,453

9.   89,375
      − 34,767

10. 32,745 − 17,676 = _____

11. 56,044 − 50,234 = _____

Problem Solving

12. William has 79,890 points on a computer game. Sarah has 58,009.

How many more points does William have than Sarah?

_____

13. Jamaica has 56,009 points, Seth has 45,334 points, and Jermaine has 5,899 points. If these players add their points together, how many points

will they have? _____

Review and Remember

Give the next three numbers in each pattern.

14. 56, 59, 62, _____

15. 230, 200, 170, _____

16. 490, 494, 498, _____

17. 1,107; 1,207; 1,307; _____

Subtracting Across Zeros

Subtract. Regroup with zeros.

1. 500
     − 490

2. 703
     − 684

3. 605
     − 360

4. 4,069
     − 1,599

5. 8,024
     − 5,899

6. 69,009
     − 52,324

7. 50,089
     − 42,866

8. 40,432
     − 15,453

9. 90,375
     − 34,353

10. 3,000
      − 1,005

11. 46,044
      − 40,234

12. 87,004
      − 25,987

Problem Solving

13. Abbey's family is working on a 1,000-piece jigsaw puzzle. There are 645 unused pieces. How many pieces has the family used in the puzzle so far?

_____

14. Pat's Restaurant has a 5,000-piece jigsaw puzzle. If there are 69 pieces remaining, how many have been used?

_____

Review and Remember

Find the missing number.

15. $7.06 − $1.49 = _____

16. 18 − ____ = 13

17. 6 × ____ = 36

18. 54 ÷ 9 = ____

19. $5.19 + $6.37 = _____

20. 72 ÷ ____ = 9

Problem Solving
Solve a Simpler Problem

Solve each problem by first solving a simpler problem.

1. Lab tables in the science room seat 4 students. Each table needs 6 test tubes to complete a lab study. How many test tubes are needed for a class of 24 students?

_____

2. It takes 20 minutes to conduct a science experiment. If each class lasts 40 minutes, how many experiments can be conducted in 5 classes?

_____

3. There are 28 students entered in the science fair. Every 4 students are allowed to have 5 visitors at the fair. How many visitors will be allowed?

_____

4. Principal Mayberry bought 48 mouse pads on sale. If the sale price was $0.50 less than the regular price, how much did he save?

_____

5. In both labs, 24 computers are set up in three equal rows. During a test, students are placed at every other station. How many students can be tested in the two labs at one time?

_____

Review and Remember

Round to the nearest ten.

6. 83 _____   **7.** 76 _____   **8.** 632 _____   **9.** 1,366 _____

Round to the nearest hundred.

10. 632 _____   **11.** 576 _____   **12.** 184 _____   **13.** 1,366 _____

Mental Math Strategies

Add or subtract mentally. Break apart the numbers.

1. 53	**2.** 73	**3.** 655	**4.** 548
− 44	+ 68	− 330	+ 523

5. 66	**6.** 868	**7.** 6,789	**8.** 5,429
− 33	+ 248	− 5,324	+ 8,645

Add or subtract mentally. Use compensation.

9. 58	**10.** 39	**11.** 76	**12.** 689
+ 42	+ 58	− 53	+ 793

13. 489	**14.** 8,453	**15.** 6,865	**16.** 2,111
− 359	− 5,899	+ 3,999	− 1,676

Problem Solving

Use mental math.

17. Julia has 13 horses and 17 dogs in her collection. How many animals does

she have altogether? _____

18. How would you solve 589 minus 365, using the compensation method?

_____

Review and Remember

Write each group of numbers in order from least to greatest.

19. 7,809 7,799 7,099 _____

20. 230,977 233,999 213,880 _____

EXPLORE: Making Change

Look at the shopping list. Write how much change you will get back for each item if you pay with a $5.00 bill.

Shopping List		
Item	**Cost**	**Change**
bread	$1.29	**1.**
milk	$3.59	**2.**
juice	$2.09	**3.**
apples	$3.45	**4.**
soup	$2.98	**5.**
crackers	$4.89	**6.**
onions	$1.01	**7.**
spaghetti	$2.39	**8.**
yogurt	$1.89	**9.**

Problem Solving

Use the table above to solve these problems.

10. How much will soup, apples, and yogurt cost? _____

11. Anna is buying crackers, milk, and bread. How much change should she get

if she pays with a $10.00 bill? _____

12. How much more expensive is milk than juice? _____

13. Which food item costs about $5.00? _____

14. Which food item costs about $1.00? _____

Review and Remember

Write the value of the underlined digit.

15. 34,809 _____   **16.** 5,979 _____

17. 877,909 _____   **18.** 230,977 _____

Choosing a Computation Method

Choose a method. Use paper and pencil, a calculator, or mental math. Write the solution.

1. 89	**2.** 79	**3.** 456	**4.** 932	**5.** 5,688	**6.** 355
+ 32	− 33	+ 365	− 65	+ 1,000	− 45

7. 8,002	**8.** 74,567	**9.** 84,909	**10.** 755	**11.** 86,909	**12.** 3,890
+ 6,500	+ 43,456	− 24,199	− 245	− 45,000	+ 4,399

Problem Solving

13. If 100 feet of wire are on one spool, and 2,000 feet are on another spool, how many feet of wire are there altogether? _____

14. If a worker needs 3,900 feet of wire for a job, does he have enough wire? How much more or less does he need? _____

15. The wire is bundled according to colors. One spool has 45 feet of green wire. Another spool has 4,009 feet of black wire. Another spool has 2,900 feet of white wire. How many feet of black and green wire are there altogether? How many more feet of black wire than white wire is there?

_____

Review and Remember

Write the standard form for each number.

16. 4 hundreds, 3 tens, 5 ones _____

17. 2 thousands, 9 tens, 2 ones _____

18. 6 hundred thousands, 7 ten thousands, 4 hundreds, 8 tens, 3 ones

_____

Problem Solving
Using Money

Many major league baseball parks have stores where visitors can buy souvenirs. Use the chart below to answer Problems 1–5.

Souvenirs	Souvenir Store's Price	Manufacturer's Cost	Money the Store Makes
T-shirts	$9.95	$4.50	?
Hats	$12.95	$5.75	?
Batting Gloves	$4.95	?	$3.50
Pennants	$2.95	$0.75	?
Baseballs	?	$1.00	$2.50

1. How much money does the store make on each hat? _____

2. What is the manufacturer's cost of batting gloves? _____

3. How much does the store make on each T-shirt? _____

4. What is the store's price for a baseball? _____

5. How much does the store make on each pennant? _____

Review and Remember

Find the answer.

6. 326
   + 141

7. 197
   + 326

8. 461
   − 131

9. 148
   − 38

10. 82
   − 45

11. 23 + 30 = _____   **12.** 67 − 50 = _____   **13.** 44 + 44 = _____

EXPLORE: Relating Multiplication and Division Facts

Complete the chart. Use counters to help you.

	Number of Buttons	Number of Friends	Drawing	Number of Buttons per Friend	Division Sentence
1.	8	4			
2.	14	2			
3.	25	5			
4.	10	5			

Problem Solving

5. Amanda has 14 party favors. She wants to divide them evenly among her

7 friends. How many party favors should she give each friend? _____

6. Josiah thinks that the division sentence for Amanda's problem is $14 \div 2 = 7$.

Do you agree or disagree? Why? _____

_____

Review and Remember

Multiply.

7. $20 \times 15 =$ _____   **8.** $4 \times 20 =$ _____   **9.** $12 \times 8 =$ _____

Multiplying by 2 and 4

Complete the first fact in each pair. Then use doubles to complete the second fact.

1. $3 \times 2 =$ _____ **2.** $3 \times 6 =$ _____ **3.** $3 \times 4 =$ _____

$6 \times 2 =$ _____ $6 \times 6 =$ _____ $6 \times 4 =$ _____

Find the product. Use doubles if you wish.

4. $\begin{array}{r} 4 \\ \times 3 \\ \hline \end{array}$ **5.** $\begin{array}{r} 4 \\ \times 8 \\ \hline \end{array}$ **6.** $\begin{array}{r} 4 \\ \times 9 \\ \hline \end{array}$ **7.** $\begin{array}{r} 2 \\ \times 5 \\ \hline \end{array}$

8. $\begin{array}{r} 7 \\ \times 2 \\ \hline \end{array}$ **9.** $\begin{array}{r} 4 \\ \times 4 \\ \hline \end{array}$ **10.** $\begin{array}{r} 7 \\ \times 6 \\ \hline \end{array}$ **11.** $\begin{array}{r} 5 \\ \times 6 \\ \hline \end{array}$

12. $\begin{array}{r} 6 \\ \times 3 \\ \hline \end{array}$ **13.** $\begin{array}{r} 8 \\ \times 2 \\ \hline \end{array}$ **14.** $\begin{array}{r} 5 \\ \times 4 \\ \hline \end{array}$ **15.** $\begin{array}{r} 6 \\ \times 2 \\ \hline \end{array}$

Problem Solving

16. Rashan wants to plant 3 rows of corn. Each row is 4 feet long. How many feet of corn will he plant? _____

17. You see a wallet for $5.97. If you wish to have 3 initials engraved on it for 25 cents each, how much will the wallet cost? _____

Review and Remember

Add or subtract.

18. $\begin{array}{r} 126 \\ + 359 \\ \hline \end{array}$ **19.** $\begin{array}{r} 423 \\ - 159 \\ \hline \end{array}$ **20.** $\begin{array}{r} 14,235 \\ + 5,234 \\ \hline \end{array}$ **21.** $\begin{array}{r} 76,525 \\ - 39,252 \\ \hline \end{array}$

Properties of Multiplication

Use properties to solve.

1. $3 \times 0 =$ _____

2. $1 \times 656 =$ _____

3. $533 \times 1 =$ _____

4. $6 \times 2 = n \times 6$

$n =$ _____

5. $n \times 8 = 8 \times 7$

$n =$ _____

6. $(n \times 5) \times 3 = 4 \times (5 \times 3)$

$n =$ _____

Find the missing number. Name the property that helps you.

7. $(n \times 2) \times 1 = 5 \times (2 \times 1)$

$n =$ _____

8. $8 \times n = 0$

$n =$ _____

Problem Solving

9. Ho is planting a vegetable garden. He has marked it into 3 rows of 4 plants in each row. His brother has marked another garden into 4 rows of 3 plants each. What is the difference in the number of plants in each garden? Explain your answer.

_____

10. Mr. Remy is teaching mathematics to his third-grade class. One group thinks that any number multiplied by 0 is that number. Another group thinks that any number multiplied by 0 is 0. What should Mr. Remy tell his class about properties? What examples could he use?

_____

_____

Review and Remember

Add or subtract.

11. $\begin{array}{r} 435 \\ -\ 129 \end{array}$

12. $\begin{array}{r} 322 \\ -\ 214 \end{array}$

13. $\begin{array}{r} \$1.42 \\ +\ 5.23 \end{array}$

14. $\begin{array}{r} \$13.95 \\ -\ 1.29 \end{array}$

Multiplying by 3 and 6

Complete the first fact. Then use doubles to complete the second fact.

1. $3 \times 8 =$ _____

$6 \times 8 =$ _____

2. $3 \times 9 =$ _____

$6 \times 9 =$ _____

3. $3 \times 3 =$ _____

$6 \times 3 =$ _____

Multiply. Use doubles if you wish.

4. $\begin{array}{r} 4 \\ \times 3 \\ \hline \end{array}$

5. $\begin{array}{r} 6 \\ \times 8 \\ \hline \end{array}$

6. $\begin{array}{r} 3 \\ \times 9 \\ \hline \end{array}$

7. $\begin{array}{r} 2 \\ \times 6 \\ \hline \end{array}$

8. $\begin{array}{r} 7 \\ \times 3 \\ \hline \end{array}$

9. $\begin{array}{r} 3 \\ \times 6 \\ \hline \end{array}$

10. $\begin{array}{r} 7 \\ \times 6 \\ \hline \end{array}$

11. $\begin{array}{r} 5 \\ \times 6 \\ \hline \end{array}$

12. $\begin{array}{r} 6 \\ \times 3 \\ \hline \end{array}$

13. $\begin{array}{r} 8 \\ \times 3 \\ \hline \end{array}$

14. $\begin{array}{r} 7 \\ \times 6 \\ \hline \end{array}$

15. $\begin{array}{r} 6 \\ \times 2 \\ \hline \end{array}$

Problem Solving

Use the table to solve Problems 16–17.

Color of Pen	Number of Pens per Package
Red	3
Blue	2
Black	6

16. How many pens will you get if you buy 6 packages of blue pens? _____

17. How many pens will you get if you buy 2 packages of each color? _____

Review and Remember

Add or subtract.

18. $\begin{array}{r} 236 \\ + 956 \\ \hline \end{array}$

19. $\begin{array}{r} 569 \\ - 199 \\ \hline \end{array}$

20. $\begin{array}{r} 4,235 \\ + 6,236 \\ \hline \end{array}$

21. $\begin{array}{r} \$7.61 \\ - 3.99 \\ \hline \end{array}$

Multiplying by 5 and 10

Find each product.

1. $7 \times 10 =$ _____   **2.** $9 \times 10 =$ _____   **3.** $4 \times 10 =$ _____

4. 10   **5.**   6   **6.**   3   **7.**   5
   $\times 3$ $\times 5$ $\times 10$ $\times 7$

8. 10   **9.**   5   **10.**   7   **11.**   5
   $\times 7$ $\times 6$ $\times 5$ $\times 8$

Compare. Use $>$, $<$, or $=$.

12. 5×9 $\bigcirc$ 9×10 **13.** 8×1 $\bigcirc$ 4×10

14. 10×2 $\bigcirc$ 5×5 **15.** 2×10 $\bigcirc$ 5×4

Problem Solving

Use the table to solve Problem 16.

Seaside Apartments	
Type of Building	Parking Spaces per Building
5-unit	5
10-unit	10

16. At Seaside apartments, there are 45 parking spaces. There are more 10-unit buildings than 5-unit buildings. How many 5-unit buildings are there? How many 10-unit buildings are there?

Review and Remember

Estimate each sum or difference.

17.   496   **18.**   599   **19.** 8,235   **20.** $9.61
   $+ 976$ $- 398$ $+ 4,936$ $- 2.99$

Multiplying by 7, 8, 9

Find each product.

1. $7 \times 10 =$ _____   **2.** $9 \times 7 =$ _____   **3.** $7 \times 7 =$ _____

4. $8 \times 5 =$ _____   **5.** $7 \times 5 =$ _____   **6.** $5 \times 9 =$ _____

Multiply.

7. $\begin{array}{r} 10 \\ \times 7 \\ \hline \end{array}$ **8.** $\begin{array}{r} 7 \\ \times 6 \\ \hline \end{array}$ **9.** $\begin{array}{r} 10 \\ \times 9 \\ \hline \end{array}$ **10.** $\begin{array}{r} 8 \\ \times 5 \\ \hline \end{array}$

Find each missing number.

11. $5 \times 7 = n$ **12.** $8 \times n = 32$ **13.** $n \times 9 = 63$ **14.** $9 \times 8 = n$

$n =$ _____ $n =$ _____ $n =$ _____ $n =$ _____

Problem Solving

Use the graph to answer Problems 15–16.

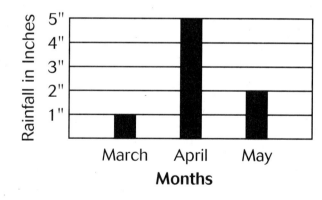

15. During which month was there the greatest rainfall? _____

16. During which two months did the rainfall total 3 inches?

_____

Review and Remember

Use a calculator or paper and pencil to add or subtract.

17. $\begin{array}{r} \$24.96 \\ + 9.27 \\ \hline \end{array}$ **18.** $\begin{array}{r} 5,099 \\ - 1,398 \\ \hline \end{array}$ **19.** $\begin{array}{r} \$8,465 \\ + \$4,976 \\ \hline \end{array}$ **20.** $\begin{array}{r} \$369 \\ - 233 \\ \hline \end{array}$

Problem Solving
Choose the Operation

Answer each question. Circle the letter of the correct answer.

Roberta is selling newspaper subscriptions. She sold 17 on Monday, 28 on Tuesday, and 12 on Wednesday. How many subscriptions has she sold so far?

1. What do you need to find?

    a. How many subscriptions she sold each day.

    b. How many more subscriptions she sold on Wednesday

    c. How many subscriptions she sold altogether

2. What do you need to do to solve the problem?

    a. Find the difference in the number sold on Monday and the number sold on Wednesday.

    b. Find the sum of subscriptions sold on all three days.

    c. Find the sum of subscriptions sold on Tuesday and Wednesday.

3. Which number sentence shows you how many subscriptions Roberta sold?

    a. $28 - 17$

    b. $28 + 12$

    c. $17 + 28 + 12$

Review and Remember

Add or subtract.

4. $6.24	**5.** $7.89	**6.** $5.19	**7.** $12.35
$+ 2.76$	$- 3.98$	$+ 2.50$	$- 5.45$

8. $3.85 + $0.75 = _____

9. $5.14 - $0.25 = _____

Relating Multiplication and Division Facts

Write the fact family for each set of numbers.
Use counters or draw arrays if you wish.

1. 5, 4, 20   **2.** 7, 9, 63   **3.** 6, 8, 48

_____   _____   _____

_____   _____   _____

_____   _____   _____

_____   _____   _____

Complete each fact family.

4. $8 \times 7 = 56$ **5.** $7 \times 6 = 42$ **6.** $10 \times 9 = 90$ **7.** $8 \times 5 = 40$

$7 \times 8 =$ ___ $6 \times 7 =$ ___ $9 \times 10 =$ ___ $5 \times 8 =$ ___

$56 \div 8 =$ ___ $42 \div 6 =$ ___ $90 \div 9 \ =$ ___ $40 \div 8 =$ ___

$56 \div 7 =$ ___ $42 \div 7 =$ ___ $90 \div 10 =$ ___ $40 \div 5 =$ ___

Problem Solving

8. Draw four possible arrays for 72 marbles.

9. Twenty-seven students are traveling to the seashore for a science field trip.
How many arrays of equal groups are possible? What are they?

_____

Review and Remember

Add, subtract, or multiply.

10.   33   **11.**   6   **12.** $5.27   **13.**   8
        $+ 54$ $\times 9$ $- 4.55$ $\times 8$

Dividing by 2 and 4

Find each quotient.

1. $6 \div 2 =$ _____
2. $16 \div 2 =$ _____
3. $4 \div 4 =$ _____
4. $24 \div 4 =$ _____

$6 \div 3 =$ _____
$16 \div 4 =$ _____
$4 \div 1 =$ _____
$24 \div 6 =$ _____

Divide.

5. $2\overline{)18}$
6. $4\overline{)16}$
7. $2\overline{)12}$
8. $4\overline{)20}$
9. $4\overline{)28}$

10. $2\overline{)14}$
11. $4\overline{)4}$
12. $2\overline{)16}$
13. $2\overline{)2}$
14. $2\overline{)6}$

Find the quotient or missing factor.

15. $18 \div 2 = n$
16. $n \div 8 = 2$
17. $24 \div n = 6$
18. $20 \div 5 = n$

$n =$ _____
$n =$ _____
$n =$ _____
$n =$ _____

19. $36 \div 9 = n$
20. $n \div 4 = 1$
21. $12 \div n = 6$
22. $n \div 2 = 7$

$n =$ _____
$n =$ _____
$n =$ _____
$n =$ _____

Problem Solving

23. A bottle manufacturer places 4 bottles in each carton. If there are 36 bottles

to be packed, then how many cartons will be filled? _____

24. Seth has 8 pencils in his collection. Two are from the computer museum, and 2 are from the science museum. The rest are from the art museum.

How many are from the art museum? _____

Review and Remember

Write the value of 3 in each number.

25. 32
26. 13
27. 13,980
28. 309
29. 375,980

_____
_____
_____
_____
_____

Dividing by 3 and 6

Find each quotient.

1. $18 \div 6 =$ _____ **2.** $27 \div 3 =$ _____ **3.** $9 \div 3 =$ _____ **4.** $24 \div 3 =$ _____

  $18 \div 3 =$ _____ $27 \div 9 =$ _____ $6 \div 3 =$ _____ $24 \div 8 =$ _____

Find each quotient.

5. $3\overline{)18}$ **6.** $3\overline{)15}$ **7.** $6\overline{)12}$ **8.** $9\overline{)27}$ **9.** $3\overline{)9}$

10. $9\overline{)18}$ **11.** $6\overline{)42}$ **12.** $6\overline{)36}$ **13.** $3\overline{)21}$ **14.** $6\overline{)6}$

Find the quotient or missing factor.

15. $18 \div 6 = n$ **16.** $n \div 3 = 3$ **17.** $24 \div n = 8$ **18.** $27 \div 3 = n$

  $n =$ _____ $n =$ _____ $n =$ _____ $n =$ _____

19. $36 \div 9 = n$ **20.** $n \div 6 = 10$ **21.** $12 \div n = 6$ **22.** $n \div 9 = 5$

  $n =$ _____ $n =$ _____ $n =$ _____ $n =$ _____

Problem Solving

23. A marathon trainer runs 24 miles in 3 hours. How many miles per hour does the trainer run? _____

24. A package of 9 pens costs $27.00. How much does each pen cost?

_____

Review and Remember

Write the number in word form.

25. 8,213 _____

26. 90,980 _____

27. 25,631 _____

Rules of Division

Use the division rules to find the missing number.

1. $0 \div 6 =$ _____   **2.** $9 \div$ ____ $= 1$ **3.** ____ $\div 1 = 18$ **4.** $81 \div$ ____ $= 1$

5. ____ $\div 7 = 0$ **6.** ____ $\div 5 = 1$ **7.** $1 \div$ ____ $= 1$ **8.** $12 \div$ ____ $= 12$

Use division rules to find the missing number. If it is not possible to divide, explain why.

9. $6 \div 6 = n$

_____

10. $7 \div n = 7$

_____

11. $n \div 0 = 18$

_____

12. $n \div 6 = 0$

_____

13. $n \div 5 = 1$

_____

14. $8 \div 0 = n$

_____

Compare. Use $>$, $<$, or $=$.

15. $7 \div 7 \bigcirc 6 \div 6$ **16.** $0 \div 7 \bigcirc 9 \div 1$ **17.** $4 \div 4 \bigcirc 0 \div 6$

18. $5 \div 5 \bigcirc 4 \div 1$ **19.** $0 \div 3 \bigcirc 0 \div 2$ **20.** $8 \div 1 \bigcirc 0 \div 8$

Problem Solving

21. Erin has four goldfish. If she divides them evenly among four friends, will she have any left for herself? _____

22. Music tapes are $1.00 at a yard sale. How many can a customer buy if he has $12.00 to spend on tapes? _____

Review and Remember

Use the numbers 5, 2, 7, 9, 3 to answer the questions.

23. Write the greatest number possible. _____

24. Write the least number possible. _____

Name _____

Dividing by 5 and 10

Find each quotient.

1. $5\overline{)25}$ **2.** $5\overline{)50}$ **3.** $10\overline{)70}$ **4.** $10\overline{)90}$ **5.** $5\overline{)40}$

6. $30 \div 10 =$ ___ **7.** $15 \div 5 =$ ___ **8.** $55 \div 5 =$ ___ **9.** $20 \div 5 =$ ___

10. $10\overline{)40}$ **11.** $10\overline{)70}$ **12.** $5\overline{)35}$ **13.** $5\overline{)15}$ **14.** $10\overline{)20}$

15. $9\overline{)27}$ **16.** $2\overline{)12}$ **17.** $4\overline{)36}$ **18.** $3\overline{)18}$ **19.** $6\overline{)54}$

20. $21 \div 3 =$ ___ **21.** $42 \div 6 =$ ___ **22.** $16 \div 2 =$ ___ **23.** $32 \div 4 =$ ___

Problem Solving

24. At Town Field Day, each team has ten players. There are 50 players. How many teams can be made? _____

25. During events at Town Field Day, spectators sit in bleachers that are stacked in 5 rows. Thirty spectators are expected for the potato sack race. How many spectators will fit into each row? _____

26. There are 45 people in the relay race. Each team has 5 runners. How many teams are there? _____

Review and Remember

Use a calculator to find n.

27. $5 \times n = 85$ **28.** $5,963 - n = 596$ **29.** $n + 3,449 = 12,384$

$n =$ _____ $n =$ _____ $n =$ _____

30. $6 \times n = 72$ **31.** $n + 2,394 = 3,018$ **32.** $6,900 - n = 3,284$

$n =$ _____ $n =$ _____ $n =$ _____

Problem Solving
Act It Out

Use counters to act out each problem.

1. Jasmine is planting tulip bulbs. She has 48 bulbs. How many ways can she arrange the bulbs so that she has the same number of tulips in each row?

_____

_____

2. Can Jasmine arrange the bulbs to form a square array? Explain.

_____

_____

3. If Jasmine was given one more bulb, could she arrange the bulbs to form a square?

_____

4. Jasmine's tulips are three different colors—red, yellow, and purple. Show at least two different arrangements for placing the different colors.

Review and Remember

Add or subtract.

5.	55	**6.**	95	**7.**	744	**8.**	684
	$+ \ 79$		$- \ 78$		$- \ 432$		$- \ 491$

Add or subtract using estimation.

9. $52 + 77$ **10.** $81 - 17$ **11.** $682 + 831$

_____   _____   _____

Dividing by 7, 8, and 9

Find each quotient.

1. 8$\overline{)48}$ **2.** 7$\overline{)49}$ **3.** 8$\overline{)16}$ **4.** 7$\overline{)63}$ **5.** 8$\overline{)56}$

6. 32 ÷ 8 = ___ **7.** 28 ÷ 7 = ___ **8.** 80 ÷ 8 = ___ **9.** 42 ÷ 7 = ___

10. 9$\overline{)81}$ **11.** 7$\overline{)70}$ **12.** 8$\overline{)32}$ **13.** 8$\overline{)16}$ **14.** 10$\overline{)20}$

15. 9$\overline{)27}$ **16.** 8$\overline{)40}$ **17.** 7$\overline{)42}$ **18.** 7$\overline{)21}$ **19.** 7$\overline{)42}$

20. 7$\overline{)21}$ **21.** 9$\overline{)9}$ **22.** 9$\overline{)36}$ **23.** 9$\overline{)18}$ **24.** 9$\overline{)54}$

Problem Solving

Complete each table.

Rule: Divide by 8

	Input	Output
25.	64	
26.		8
27.	56	
28.		2
29.	24	

30. Rule: _____

Input	Output
63	7
45	5
18	2
27	3
36	4

Review and Remember

Choose +, −, ×, or ÷ to make each sentence true.

31. 5 $\bigcirc$ 5 = 1 **32.** 9 $\bigcirc$ 3 = 27 **33.** 1,643 $\bigcirc$ 345 = 1,988

Use with Grade 4, text pages 116-119. **37**

Problem Solving
Using a Pictograph

Corey did a survey of students. He asked each student to name their favorite school subject. The pictograph below shows the results. Use the graph to answer Problems 1–5.

Favorite School Subjects

Reading	Science	Math	Computers	History	Art
			🚶 🚶		
		🚶	🚶 🚶		
		🚶 🚶	🚶 🚶		
	🚶 🚶	🚶 🚶	🚶 🚶		🚶 🚶
🚶 🚶	🚶 🚶	🚶 🚶	🚶 🚶		🚶 🚶
🚶 🚶	🚶 🚶	🚶 🚶	🚶 🚶	🚶	🚶 🚶
🚶 🚶	🚶 🚶	🚶 🚶	🚶 🚶	🚶 🚶	🚶 🚶
🚶 🚶	🚶 🚶	🚶 🚶	🚶 🚶	🚶 🚶	🚶 🚶

Each 🚶 stands for 5 students

1. How many students answered the survey? _____

2. How many different subjects were named? _____

3. What subject seems to be most popular? _____

4. What subject seems to be least popular? _____

5. How many students named math as their favorite? _____

Review and Remember

Multiply.

6. $6 \times 5 =$ ____   **7.** $9 \times 4 =$ ____   **8.** $4 \times 8 =$ ____   **9.** $8 \times 5 =$ ____

Find the missing factor.

10. $4 \times$ ____ $= 20$ **11.** $3 \times$ ____ $= 24$ **12.** ____ $\times 5 = 30$ **13.** ____ $\times 7 = 28$

Explore: Collecting and Organizing Data

Use the table to answer these problems.

Lakeview Hornets Basketball Team		
Player	**Grade**	**Points**
Jonathon	Grade 5	80
Katy	Grade 5	72
Shateel	Grade 4	23
Jamal	Grade 3	63
Jennifer	Grade 4	14
Seth	Grade 5	55
Kristin	Grade 3	32

1. How many players are included in the table? _____

2. Who scored the most points in Grade 4? _____

3. What is the median of the list of points? _____

4. Who scored the fewest points? _____

Problem Solving

5. Seventeen players are listed on a score card. More players have a score of 14 than any other score. Seamus thinks 14 is the median. Alex thinks 14 is the mode. With whom do you agree? Why? _____

_____

6. Karyn wants to buy a jacket for $44.99. She has saved $23.00. If she babysits for 5 hours and earns $3.25 per hour, will she then have enough money to

buy the coat? _____

Review and Remember

Solve.

7. $4 \times 8 =$ _____   **8.** $81 \div 9 =$ _____   **9.** $10 \times 12 =$ _____

Use with Grade 4, text pages 134–137.   **39**

Making a Bar Graph

1. Make a bar graph for the data in the table.

Number of Pets	
Student	**Pet**
Tim	dog
Kyle	dog
Jose	cat
Yuneng	bird
Heather	fish
Amanda	cat
Eva	dog

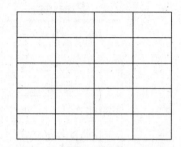

2. Which is the least common pet? _____

3. Which is the most common pet? _____

4. Who has a bird as a pet? _____

Problem Solving

Use the graph to solve Problems 5–6.

5. How many inches taller is the daisy than the tulip?

_____

6. Which plant is the tallest?

_____

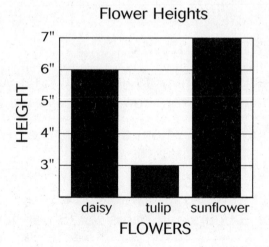

Flower Heights

Review and Remember

Solve.

7. 7,235 + 563 = _____

8. 9,073 − 1,245 = _____

Making Pictographs

Use the pictograph to answer Problems 1–3.

1. Which team scored the most points?

_____

2. What is the total number of points

scored by all four teams? _____

3. How many more points did the

Redwings score than the Bluebirds?

_____

Number of Points Scored	
Archery Team	**Points**
Robins	→ → → →
Parrots	→ → → → → → → →
Bluebirds	→ → → → →
Redwings	→ → → → → →

Each → stands for 4 points.

Problem Solving

Make a pictograph for the data in the table.

Bowling Tournament	
Games Won	**Player**
4	Pietro
3	Chelsea
2	Barry
4	Olin
5	Mary Jo

4. Which two players are tied for second place? _____

5. Did you need to use a half symbol to make the pictograph?

Review and Remember

Use a calculator to add or subtract.

6. 17,235 + 5,963 = _____

7. 61,369 − 5,299 = _____

Problem Solving
Understanding Line Graphs

Use the graph at the right to answer Problems 1–3.
Circle the letter of the correct answer.

1. Which of the following is true?

  a. Judy finished the trip at point *F*.

  b. The trip is half over at point *C*.

  c. You cannot tell where the trip ended.

2. Which statement best describes what happened between points *A* and *C* on the graph?

  a. Judy drove some distance each day.

  b. Judy drove the same distance each day.

  c. Judy stopped for gas each day.

3. Which statement best describes what happened between points *C* and *D* on the graph?

  a. Judy continued her trip.

  b. Judy did not travel.

  c. Judy turned toward home.

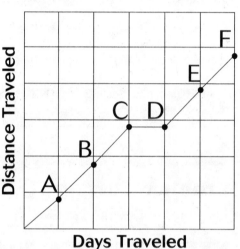

Judy's Cross-Country Trip

Distance Traveled / Days Traveled

Review and Remember

Multiply.

4.   3
  × 4

5.   3
  × 8

6.   2
  × 9

7.   9
  × 3

8.   7
  × 7

9.   6
  × 5

10.   7
  × 4

11.   5
  × 8

12.   8
  × 9

13.   9
  × 6

14.   8
  × 7

15.   4
  × 5

Reading Line Graphs

Use the line graph to answer these problems.

1. During which hour did Rachel bike the farthest?

_____

2. Did Rachel bike farther between 10:00 A.M. and 11:00 A.M. or between 3:00 P.M. and 4:00 P.M.? Tell how you know.

_____

_____

_____

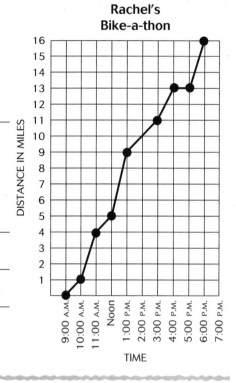

Rachel's Bike-a-thon

Problem Solving

Solve.

3. Rachel collects $5.00 for each mile she bikes. How much will she collect altogether? _____

4. If Rachel collects only $4.00 for each mile she bikes, then how much less will she collect? _____

Review and Remember

Find each missing number.

5. $n \times 60 = 12 \times 10$

$n =$ _____

6. $45 - n = 50 - 10$

$n =$ _____

7. $30 + 10 = 2 \times n$

$n =$ _____

8. $0 \times 81 = n \times 5$

$n =$ _____

Graphing Ordered Pairs

Use the grid below. Write the ordered pair for each point.

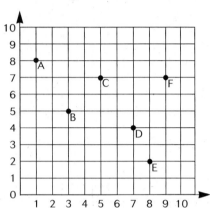

1. point A _____

2. point D _____

3. point F _____

4. point C _____

Problem Solving

Use the grid to answer Problems 5–6.

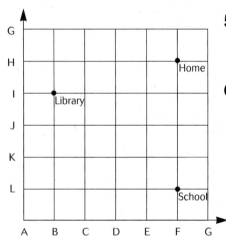

5. What point is halfway between home and school?

_____

6. Suppose you wanted to go the shortest distance from the library to home. You can only follow the lines on the grid. Through which points would you pass?

_____

_____

Review and Remember

Find each product or quotient.

7. $8 \times 9 =$ _____

8. $64 \div 8 =$ _____

9. $7 \times 6 =$ _____

10. $45 \div 5 =$ _____

11. $2 \times 10 =$ _____

12. $36 \div 6 =$ _____

13. $8 \times 7 =$ _____

14. $27 \div 9 =$ _____

15. $9 \times 7 =$ _____

16. $81 \div 3 =$ _____

17. $8 \times 8 =$ _____

18. $25 \div 5 =$ _____

Problem Solving
Make a Graph

The chart shows the results of a survey on
school lunch choices.
Make a graph to display the data given.
Use the graph to answer Problems 1–3.

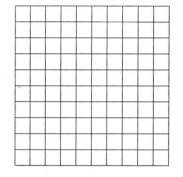

Lunch Favorite		
Pizza	**Soup and Sandwich**	**Pasta**
‖‖ ‖‖ ‖‖ ‖‖ ‖‖ ‖‖ ‖‖	‖‖ ‖‖ ‖‖ ‖‖ ‖‖ ‖‖ ‖‖ ‖‖ ‖‖ |||	‖‖ ‖‖ ‖‖ ‖‖ ‖‖ ‖‖ ‖‖ ||

1. What school lunch is the most popular among the students?

_____

2. Compare the number of votes for pizza and pasta.

_____

3. What would the graph show if pizza got 10 more votes?

_____

4. The cafeteria serves lunch to about 12 teachers each day.
How many teachers do they serve each week?

_____

Review and Remember

Write the place value of the underlined digit.

5. 5,368 _____

6. 5_7_5 _____

7. _7_62 _____

Compare. Write >, <, or =.

8. 7 + 4 ◯ 6 + 5

9. 8 + 4 ◯ 9 + 2

10. 9 − 5 ◯ 7 − 2

Reading Stem-and-Leaf Plots

Use this stem-and-leaf plot for Problems 1–4.

**Points Earned
by Teams A and B**

Stem	Leaf
2	5, 8, 8, 8, 9
3	0, 1, 5, 6, 8, 9

1. What was the least number of points scored? _____

2. What was the greatest number of points scored? _____

3. What is the mode? _____

4. How many games did the teams play? _____

Problem Solving

5. Each time Teams A and B scored 28 points, they won the match. How many

times did they win? _____

6. For each point scored over 34, teams got an extra turn. How many extra turns

did Teams A and B get? _____

Review and Remember

Write the fact family for each set of numbers.

7. 3, 7, 21

_____

8. 6, 4, 24

_____

Problem Solving
Ways to Represent Data

Solve each problem. Tell whether you used the pictograph or the bar graph. Explain your choice.

	Team Wins
Sept	⚾⚾⚾⚾⚾⚾
August	⚾⚾⚾⚾⚾⚾⚾⚾⚾⚾⚾⚾
July	⚾⚾⚾⚾⚾⚾
June	⚾⚾⚾⚾
May	⚾⚾⚾⚾
Apr	⚾⚾

Each ⚾ stands for 2 wins

Team Wins

(bar graph: Wins vs Month — Apr, May, Jun, Jul, Aug, Sept)

1. In what months did the team play their games?

_____

2. How many games did the team win in July? _____

3. In what month did the team win the fewest games? _____

4. In what months did the team win 10 or more games?

_____

Review and Remember

Add or subtract.

5. 145
      + 69

6. 148
      + 507

7. 240
      + 178

8. 187
      + 98

9. 754
      −103

10. 957
       − 138

11. 100
       − 64

12. 805
       − 126

Name _____

Understanding Probability

Describe how likely it is that each event will happen. Use the words *impossible, not likely, very likely,* and *certain.*

1. All of the trees in the park will grow. _____

2. All of the trees in the park will die someday. _____

3. All of the trees in the park have tree trunks. _____

4. All of the trees in the park will grow a mile high. _____

Problem Solving

Write either *more likely* or *less likely* in each blank.

5. In the city park it is _____ that sparrows will nest in

the trees. It is _____ that eagles will nest in the trees.

6. It is _____ that a tornado will damage the trees. It is

_____ that gentle wind will damage the trees.

Review and Remember

Write the word names for these numbers.

7. 717 _____

8. 9,099 _____

9. 575,397 _____

_____

10. 1,307,512 _____

_____

EXPLORE: Listing Outcomes

There are 6 different shapes in a bag. Each student will draw one shape, trace it, then return it to the bag.

1. How many outcomes are possible? _____

2. Are all of the outcomes equally likely? Tell why or why not.

_____

_____

3. Would the outcomes change if each student took a shape and did not replace

it in the bag? Why or why not? _____

_____

_____

Problem Solving

Use the spinner to the right to answer Problems 4–5.

4. If you spin the spinner once, how many outcomes are

possible? Name each outcome. _____

5. Are all the outcomes of spinning the spinner equally likely? Explain why or

why not. _____

_____

Review and Remember

Write in words the value of the underlined digit.

6. 7,2̲17 _____

7. 23̲,904 _____

8. 5̲75,397 _____

EXPLORE: Evaluating Fairness

Which of these spinners would be fair? Which would be unfair? Tell why.

1.

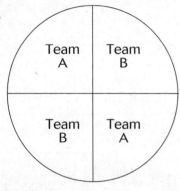

2.

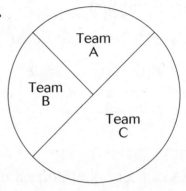

_____   _____

_____   _____

Problem Solving

3. Jason thinks that a game with equally likely outcomes is unfair because the points on the spinner are odd numbers. Do you agree? Why or why not?

_____

_____

4. Kathryn designed a board game. Players begin with 5 points and get the same number of spins. The first player to reach 100 points is the winner. Do you think the game is fair? Tell why or why not.

_____

Review and Remember

Compare. Write <, >, or =.

5. 890 ◯ 889   **6.** 72 ÷ 9 ◯ 54 ÷ 9   **7.** 18 × 2 ◯ 72 ÷ 2

Mental Math: Multiplying Multiples of 10, 100, and 1,000

Find the product.

1. $7 \times 2 =$ _____

$7 \times 20 =$ _____

$7 \times 200 =$ _____

$7 \times 2,000 =$ _____

2. $5 \times 3 =$ _____

$5 \times 30 =$ _____

$5 \times 300 =$ _____

$5 \times 3,000 =$ _____

3. $6 \times 2 =$ _____

$6 \times 20 =$ _____

$6 \times 200 =$ _____

$6 \times 2,000 =$ _____

Use basic facts and patterns to find each product.

4. $3 \times 20 =$ _____

5. $5 \times 400 =$ _____

6. $9 \times 300 =$ _____

7. $6 \times 80 =$ _____

8. $3 \times 3,000 =$ _____

9. $4 \times 400 =$ _____

Problem Solving

10. Jonnie thinks that 3×300 is 900. JoBeth thinks that 3×300 is 9,000. With whom do you agree? Why?

_____

11. If Sandy reads 100 pages in one hour, then how many pages will she read

in 10 hours? _____

Review and Remember

Write the fact family for each set of numbers.

12. 4, 5, 20

13. 2, 8, 16

14. 3, 2, 6

_____   _____   _____

_____   _____   _____

_____   _____   _____

_____   _____   _____

Estimating Products

Round to the underlined place. Estimate the product.

1. 2̲8
× 3 ____

2. 2̲2
× 4 ____

3. 2̲12
× 2 _____

4. 5̲92
× 5 _____

5. 2̲,320
× 4 _____

6. 6̲30
× 2 _____

7. 3̲,462
× 8 _____

8. 2̲93
× 9 _____

9. 2 × $8̲63 _____

10. 8̲22 × 6 _____

11. 2̲2 × $6 _____

Choose two factors from the box for each of these estimated products.

9	12	630	763

12. 90

13. 5,400

14. 7,200

15. 9,600

_____ _____ _____ _____

Problem Solving

Use the chart to answer questions below.

16. About how many hours will

David use in 3 years? _____

17. About how many hours did

the people use altogether? _____

Internet Hours per Year	
David	123
Yuneng	29
Keith	465
Crystal	52
Devon	964

Review and Remember

Estimate the sum or difference.

18. 29 + 48

19. 4,520 + 1,146

20. 2,816 − 101

21. 326 − 202

_____ _____ _____ _____

Problem Solving
Reasonable Answers

Answer each question. Give a reason for your choice.

Pedro's store receives 8 cartons of piñatas for the holiday. Each carton has 18 piñatas. He sells 4 cartons of piñatas to a school.

1. Which of the following could you use to estimate the number of piñatas Pedro has left after the sale?

    a. 8×18 **b.** 8×20 **c.** 4×18

2. Would it be reasonable to say that the school receives about 80 piñatas? Why or why not?

_____

_____

Maria is having a party with 42 guests. She wants each guest to have 3 party favors each. Party favors come in bags of 8.

3. Which of the following could you use to estimate the number of favors that are needed for the guests?

    a. 40×3 **b.** $42 \div 8$ **c.** 40×8

4. Would it be reasonable to say that 10 bags of party favors are needed for the guests? Why or why not?

_____

_____

Review and Remember

Round each number to the underlined place.

5. 8<u>2</u> _____   **6.** <u>5</u>73 _____   **7.** 3,<u>4</u>56 _____

8. 4<u>5</u>9 _____   **9.** <u>4</u>,803 _____   **10.** 5,<u>0</u>19 _____

EXPLORE: Multiplication by One-Digit Numbers

Round to the underlined place. Estimate the product. Then multiply.

1. $\underline{6}8$
$\times\ 3$ _____

2. $\underline{7}2$
$\times\ 7$ _____

3. $\underline{7}12$
$\times\ 6$ _____

4. $\underline{6}42$
$\times\ 5$ _____

5. $\underline{2},140$
$\times\ 4$ _____

6. $\underline{5}30$
$\times\ 2$ _____

7. $\underline{6},461$
$\times\ 8$ _____

8. $\underline{5}03$
$\times\ 9$ _____

9. $7 \times \$\underline{4}63 =$ _____

10. $\underline{5}22 \times 3 =$ _____

11. $6 \times \underline{6},632 =$ _____

12. $\underline{8}41 \times 4 =$ _____

Use base-ten blocks to model the problem. Then write the product.

13. 13
$\times\ 4$

14. 12
$\times\ 5$

15. 30
$\times\ 6$

16. 46
$\times\ 2$

Problem Solving

17. If one box of blocks costs $17.98, then about how much would 5 boxes of blocks cost? _____

18. 7 times a number is 42,000. What is the number? _____

Review and Remember

Find the answer.

19. 37
$+\ 29$

20. 520
$+\ 963$

21. 863
$-\ 795$

22. 6,952
$-\ 3,965$

Multiplying Two-Digit Numbers

Estimate. Then find the product. Use base-ten blocks to help you.

1. 55
  × 4 _____

2. 23
  × 7 _____

3. 74
  × 6 _____

4. 28
  × 2 _____

5. 34
  × 3 _____

6. 15
  × 5 _____

7. 44
  × 8 _____

8. 92
  × 9 _____

9. $7 \times 46 =$ _____ **10.** $57 \times 3 =$ _____ **11.** $82 \times 9 =$ _____

Find the product. Use base-ten blocks to help you.

12. 26
  × 3

13. 15
  × 2

14. 35
  × 4

15. 59
  × 5

16. 18
  × 6

17. 72
  × 7

18. 64
  × 8

19. 48
  × 9

Problem Solving

Use the graph for Problems 20–21.

20. How many miles did Amanda and
Alex bike altogether? _____

21. How many more miles did Carter bike
than Amanda? _____

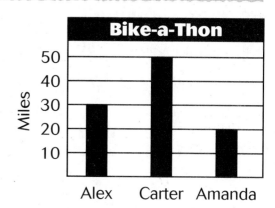

Review and Remember

Find the answer.

22. $36 \div 6 =$ ____ **23.** $54 \div 9 =$ ____ **24.** $8 \times 6 =$ ____ **25.** $7 \times 4 =$ ____

Use with Grade 4, text pages 188–191. **55**

Multiplying Two-Digit Numbers

Estimate. Then find the exact product.

1. 16
 × 4 _____

2. 49
 × 2 _____

3. 86
 × 5 _____

4. 35
 × 6 _____

5. 94
 × 3 _____

6. 25
 × 7 _____

7. 74
 × 8 _____

8. 16
 × 9 _____

9. $7 \times \$26 =$ _____

10. $26 \times 3 =$ _____

Find each product.

11. 16
 × 2

12. 75
 × 4

13. 33
 × 9

14. 29
 × 6

15. 43
 × 6

16. 52
 × 5

17. 34
 × 8

18. 88
 × 3

Problem Solving

19. Seamus can type 36 words per minute. His brother can type twice as many.

How many words per minute does his brother type? _____

20. Kyle wants to make paper flowers. She has 89 sheets of paper. Each flower takes 3 sheets of paper. Does she have enough paper to make 24 flowers?

_____

Review and Remember

Find the missing number.

21. $6 \div n = 3$

22. $n \div 9 = 3$

23. $4 \times n = 32$

24. $56 \div n = 7$

$n =$ _____

$n =$ _____

$n =$ _____

$n =$ _____

Problem Solving
Make a Table

Students at the Pine Hill School held a Math Olympics. For every 2 games it won, a team earned 5 points and received 1 prize. Complete the table below, which has been started for you. Then use it to solve Problems 1–4.

Pine Hill School Math Olympics		
Number of Prizes	Games Won	Number of Points
1	2	5
	4	10
3	6	
4		20
	12	

1. How many prizes will a team with 30 points earn? _____

2. The Olympic Measurement Team won 30 games. How many prizes should they claim? _____

3. The Olympic Fractions Team earned 60 points. How many prizes should they claim? _____

4. The Basic Facts team claimed 18 prizes. How many games did they win? How many points did they earn? _____

Review and Remember

Give the value of the underlined digit.

5. 6,234   **6.** 99   **7.** 890   **8.** 43   **9.** 254   **10.** 17,867

_____   _____   _____   _____   _____

Multiplying Greater Numbers

Estimate first. Then multiply to find the product.

1. 106
× 4 _____

2. 230
× 2 _____

3. 3,260
× 8 _____

4. 305
× 3 _____

5. 6,220
× 3 _____

6. 2,345
× 7 _____

7. 740
× 8 _____

8. 1,536
× 9 _____

Multiply. Use paper and pencil or a calculator. Tell which method you used.

9. 723
× 6

10. 50
× 4

11. 4,362
× 9

12. 52
× 7

Solve for *n*.

13. $4 \times 316 = n$

n = _____

14. $7 \times 524 = n$

n = _____

15. $9 \times 200 = n$

n = _____

Problem Solving

Use the pictograph to solve Problems 16–17.

16. How many books did Evon read? _____

17. How many more science fiction books than history books did Evon read?

_____

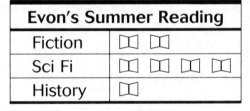

Evon's Summer Reading	
Fiction	📖 📖
Sci Fi	📖 📖 📖 📖
History	📖

📖 = 5 books

Review and Remember

Find the the missing number.

18. $6,532 - 2,369 = n$

n = _____

19. $9,522 + 362 = n$

n = _____

20. $27 \div 3 = n$

n = _____

Multiplying with Money

Estimate each answer. Then multiply to find each product.

1. $1.24
  × 3 _____

2. $8.30
  × 5 _____

3. $2.60
  × 6 _____

4. $5.36
  × 8 _____

5. $14.67
  × 9 _____

6. $21.35
  × 4 _____

7. $9.46
  × 7 _____

8. $32.97
  × 2 _____

Find each product.

9. $23.45
  × 6

10. $32.12
  × 4

11. $71.10
  × 9

12. $2.32
  × 7

Problem Solving

Use the table to answer the questions.

13. How much do 4 shirts cost if they are on sale? How much do you save if you buy them on sale?

_____

14. Ms. Cooper bought 3 of each item on sale. How much did she spend?

_____

Item	Regular Price	Sale Price
coat	$54.00	$27.89
shirt	$14.99	$12.00
jeans	$34.00	$24.99

Review and Remember

Solve for n.

15. $9.03 + n = $23.89

$n =$ _____

16. $32 \div n = $8

$n =$ _____

17. $270.69 - 32.75 = n

$n =$ _____

Problem Solving
Using Money

Use the data from the table to help you solve Problems 1–4.

Computer Technology Science Movie Prices		
Movie	Price per ticket	Package of 10 tickets
Animal Kingdom	$0.50	$4.50
Mountain Views	$1.00	$9.00
Thunderstorm!	$0.75	$7.50
Under the Sea	$1.25	$10.00

1. Which movie does not offer a savings in the per ticket cost if a customer buys a package of 10 tickets?

_____

2. Eleven students in Mr. Dumont's class are seeing *Under the Sea* and 5 students are seeing *Mountain Views*. How much will all the tickets cost?

_____

3. Ms. Wong's entire class is seeing *Animal Kingdom* as part of their study of Africa. How much will the 25 students, Ms. Wong, and 4 parents pay for admission?

_____

4. One group of six students each bought a glass of juice for 25 cents. How much change did they get from $5.00?

_____

Review and Remember

Compare. Write $>$, $<$, or $=$.

5. 565 $\bigcirc$ 56   **6.** 199 $\bigcirc$ 299   **7.** 34 $\bigcirc$ 340

Mental Math: Multiplying Multiples of Ten

Use patterns with zeros to solve.

1. $7 \times 20 =$ _____

$70 \times 20 =$ _____

$70 \times 200 =$ _____

2. $5 \times 30 =$ _____

$50 \times 30 =$ _____

$50 \times 300 =$ _____

3. $6 \times 20 =$ _____

$60 \times 20 =$ _____

$60 \times 200 =$ _____

4. $80 \times 30 =$ _____

5. $60 \times 300 =$ _____

6. $30 \times 90 =$ _____

7.   30
   $\times\ 20$

8.   500
   $\times\ 40$

9.   900
   $\times\ 30$

10.   8,000
   $\times\ \ \ 60$

11. 600
   $\times\ 80$

12.   3,000
   $\times\ \ \ 30$

13.   400
   $\times 400$

14. 7,000
   $\times\ \ \ 90$

Problem Solving

15. Forty band members will each sell 30 tickets for a concert.

How many tickets will they sell altogether? _____

16. Four hundred people bought concert tickets in advance for $10 each. Twice as many people bought tickets at the door for $20. How much money was

collected in ticket sales? _____

Review and Remember

Find each answer.

17. $4 \times 8 =$ _____

18. $49 \div 7 =$ _____

19. $63 \div 9 =$ _____

20. $709 - 635 =$ _____

21. $15 + 203 + 8 + 94 =$ _____

22. $6 \times 6 =$ _____

Estimating Products

Round each factor to the underlined place. Then estimate each product.

1. <u>7</u>2 × <u>2</u>6

___ × ___

___

2. <u>5</u>3 × <u>3</u>2

___ × ___

___

3. <u>1</u>6 × <u>2</u>1

___ × ___

___

4. <u>4</u>61 × <u>5</u>7

___ × ___

___

5. <u>8</u>30 × <u>3</u>2

___ × ___

___

6. <u>2</u>60 × <u>3</u>9

___ × ___

___

7. <u>3</u>0 × <u>9</u>20

___ × ___

___

8. <u>2</u>1 × <u>2</u>36

___ × ___

___

9. <u>6</u>,145 × <u>8</u>3

___ × ___

___

10. <u>3</u>,439 × <u>4</u>6

___ × ___

___

11. <u>4</u>,200 × <u>7</u>1

___ × ___

___

12. <u>7</u>,312 × <u>9</u>1

___ × ___

___

Problem Solving

Use the table to solve Problems 13–14.

Vehicle	Miles per Hour
Car	56
Bus	42
Rocket ship	895
Train	86
Plane	232

13. Estimate how many miles the rocket ship will travel in 21 hours.

_____

14. Estimate how many miles the car, bus, and train will each travel in

11 hours. car _____

bus _____   train _____

Review and Remember

Find the answers.

15. $8.14
         − 4.88

16.   65,952
        + 63,125

17. 3)‾27‾

18.   52
         × 6

Problem Solving

Too Much or Too Little Information

Answer each question. Give a reason for your choice.

Jennifer raises 10 chickens. She sells eggs for $1.00 per dozen. Chicken feed costs about $13.00 each month. How many dozen eggs does she need to sell to pay for the chicken feed?

1. What information is not needed to solve the problem?

    a. Jennifer raises 10 chickens.

    b. Chicken feed costs $13.00.

    c. Jennifer sells eggs for $1.00 per dozen.

2. What information is needed to solve the problem?

    a. Chicken feed costs $13.00.

    b. Jennifer sells eggs for $1.00 per dozen.

    c. Both A and B

3. Which number sentence tells how many dozen eggs Jennifer needs to sell?

    a. $12 \times 10 = 120$

    b. $13 \times \$1.00 = \$13$

    c. $10 \times \$1.00 = \$10$

Review and Remember

Round each factor to the underlined place. Then estimate each product.

4. $\underline{2}7 \times \underline{1}2 = $ _____

5. $\underline{4}4 \times \underline{3}4 = $ _____

6. $\underline{8}5 \times \underline{6}2 = $ _____

7. $\underline{8}1 \times \underline{2}3 = $ _____

8. $\underline{1}8 \times \underline{3}7 = $ _____

9. $\underline{9}0 \times \underline{7}2 = $ _____

10. $\underline{1}9 \times \underline{9}9 = $ _____

11. $\underline{4}3 \times \underline{8}7 = $ _____

12. $\underline{2}5 \times \underline{2}5 = $ _____

EXPLORE: Two-Digit Factors

Write a multiplication sentence for each rectangle in the grid below. Then add the numbers to find the total product.

1. A: _____ × _____ = _____

    B: _____ × _____ = _____

    C: _____ × _____ = _____

    D: _____ × _____ = _____

    $A + B + C + D =$ _____

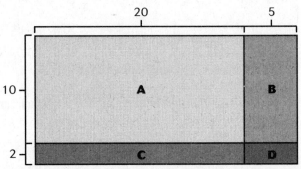

Use a grid to find each product. Divide your grid paper into smaller rectangles.

2. $13 \times 16 =$ _____

3. $11 \times 32 =$ _____

4. $24 \times 19 =$ _____

5. $21 \times 47 =$ _____

Problem Solving

6. A movie theater has 26 rows of 14 seats. How many seats are there

altogether? _____

7. Chaan buys 13 flowers for 15¢ each. He pays with a $5 bill. How much

change should he get? _____

Review and Remember

Find the answers.

8. $4 \times 8 =$ _____

9. $6 \times 7 =$ _____

10. $9 \times 8 =$ _____

11. $3 \times 7 =$ _____

12. $4 \times 7 =$ _____

13. $9 \times 9 =$ _____

14. $6 \times 2 =$ _____

15. $7 \times 8 =$ _____

Multiplying by a Multiple of Ten

Estimate first. Then multiply to find each product.

1. 30
$\times$ 80 _____

2. 63
$\times$ 10 _____

3. 2,360
$\times$ 40 _____

4. 820
$\times$ 70 _____

Find each product.

5. 12
$\times$ 50

6. 42
$\times$ 40

7. 51
$\times$ 60

8. 92
$\times$ 20

9. 123
$\times$ 30

10. 381
$\times$ 80

11. 550
$\times$ 70

12. 3,211
$\times$ 20

13. 959
$\times$ 90

14. 6,781
$\times$ 40

Problem Solving

15. How far will a car traveling 55 miles per hour travel in 10 hours?

_____

16. Tom is building a bookcase that has four shelves. Each shelf will be 16 inches

long. How many inches of shelving does he need? _____

Review and Remember

Find each missing number.

17. $421 + n = 623$

$n =$ _____

18. $\$11.36 - n = \$8.21$

$n =$ _____

19. $n - 234 = 823$

$n =$ _____

20. $7 \times n = 42$

$n =$ _____

21. $n + 85 = 212$

$n =$ _____

22. $63 \div n = 9$

$n =$ _____

Multiplying by Two-Digit Numbers

Estimate first. Then multiply to find each product.

1. 23
  × 61 _____

2. 43
  × 25 _____

3. 79
  × 34 _____

4. 27
  × 76 _____

Find each product.

5. 12
  × 52

6. 32
  × 49

7. 51
  × 68

8. 95
  × 29

9. 23
  × 38

10. 82
  × 63

11. 69
  × 43

12. 84
  × 67

13. 23
  × 63

14. 55
  × 74

Find the missing digits in each factor. Use a calculator to help you.

15. 5■ × 6■ = 3,276

16. 1■ × 1■ = 160

17. 4■ × 1■ = 611

_____   _____   _____

Problem Solving

18. Sally's dog eats 16 pounds of dog food each month. Does her dog eat more than 200 pounds in one year? _____

19. Sally's rabbit eats about 56 carrots each month. Does her rabbit eat more than 600 carrots a year? _____

Review and Remember

Use paper and pencil or a calculator to add.

20.   1,421
  90,007
  + 5,639

21. 22,562
  31,526
  +   230

22. $231.26
  963.31
  +   2.36

Problem Solving
Make a List

Use a tree diagram to make a list to solve each problem.

1. In a science experiment, students can choose to use paper towel or cloth. They must also choose either red, green, or yellow food coloring. List all the possible combinations that students can make.

2. At the school cafeteria, students can choose one beverage, one entree, and one dessert. The choices are juice or milk; hamburger or pizza; and banana or apple. List all the combinations students can make.

Review and Remember

Add or subtract.

3. $2,568 + 54 =$ _____

4. $890 - 32 =$ _____

5. $3,098 - 340 =$ _____

6. $1,400 - 97 =$ _____

7. $516 + 974 =$ _____

8. $603 + 199 =$ _____

Multiplying Greater Numbers

Estimate first. Then multiply to find each product.

1. 233	**2.** 943	**3.** 719	**4.** 207
× 41 _____	× 25 _____	× 64 _____	× 76 _____

_____

Find each product.

5. 162	**6.** 392	**7.** 518	**8.** 925	**9.** 123
× 52	× 47	× 68	× 21	× 32

10. 882	**11.** 697	**12.** 804	**13.** 423	**14.** 755
× 88	× 73	× 67	× 95	× 14

Write < , >, or = for each.

15. 52 × 6,463 ◯ 52 × 5,463   **16.** 27 × 8,692 ◯ 28 × 7,523

Problem Solving

17. Books cost $1.35 each at a yard sale. Is $20.00 enough to buy 15 books?

_____

18. Beads cost $0.65 per bag. Is $10.00 enough to buy 32 bags?

_____

Review and Remember

Use mental math to solve.

19. 525 + 25 = _____   **20.** 600 × 40 = _____   **21.** 91 − 61 = _____

22. 90 × 70 = _____   **23.** 125 − 105 = _____   **24.** 250 + 750 = _____

Problem Solving
Choose a Computation Method

Solve each problem, using paper and pencil, mental math, or a calculator. Tell which method you chose.

1. Robby's aunt is three times as old as he is. If Robby is 10, how old is

his aunt? _____

2. Tim has $5.60. How much more money does he need to buy a CD for

$16.97? _____

3. Twenty thousand people are expected to attend the mayor's inauguration. If it rains, there are seats indoors for 12,000. How many people may have

to stand in the back of the auditorium? _____

4. Fourth-grade musicians are fund-raising for new uniforms. Their earnings each week have been $14.98, $35.67, $98.34, and $56.21. How much

more do they have to raise to meet their goal of $250? _____

5. The first-place runner in one city's marathon finished in 2 hours 32 minutes. The last runner crossed the finish line at 5 hours 48 minutes. For how long

were runners crossing the finish line? _____

6. Tobias is walking in a 12-mile walk for charity. He has pledges of $2.00, $0.25, $0.75, and $1.50 per mile. One sponsor will give him $25 if he

finishes the walk. How much will Tobias collect if he finishes? _____

Review and Remember

Solve.

7. $8 \times 9 =$ _____   **8.** $12 \div 6 =$ _____   **9.** $54 \div 9 =$ _____   **10.** $6 \times 7 =$ _____

11. $63 \div 9 =$ _____   **12.** $8 \times 5 =$ _____   **13.** $6 \times 8 =$ _____   **14.** $49 \div 7 =$ _____

Telling Time

Write each missing number.

1. 72 h = _____ days   **2.** 30 min = _____ sec

3. 2 h 3 min = _____ min   **4.** 1 h 40 min = _____ min

Write each time two ways, using words and numbers.

5.

6.

_____   _____

Problem Solving

7. You are twenty minutes late for a movie that begins at 2:00. What time do you arrive at the theater? _____

8. Each commercial on television lasts for one minute. There are five commercials. How many seconds do they last? _____

Review and Remember

Solve.

9.   14
       × 8

10.   490
        − 78

11.   946
        + 63

12. 5)500

13.   531
        −209

14.   81
        × 3

15.   1,302
        − 989

16.   701
        × 10

Elapsed Time

Find each elapsed time.

1. Start 4:15 A.M.

Finish 5:30 A.M.

_____

2. Start 9:30 P.M.

Finish 1:30 A.M.

_____

3. Start 10:15 A.M.

Finish 3:30 P.M.

_____

4. Start 6:30 A.M.

Finish 1:45 P.M.

_____

Find each missing time.

	Start	Elapsed Time	Finish
5.		4 hours	10:00 A.M.
6.	7:30 P.M.		10:00 P.M.
7.	9:30 A.M.	5 hours 15 minutes	
8.	2:40 P.M.	9 hours 5 minutes	
9.	5:15 A.M.		5:15 P.M.

Problem Solving

10. Your train leaves at 7:15 A.M. and arrives at 11:40 A.M. How long is the

train ride? _____

11. If you bike for two hours and ten minutes and arrive at your destination at

1:20 P.M., then what time did you leave? _____

Review and Remember

Solve for *n*.

12. $n \times 8 = 64$

$n =$ _____

13. $369 - n = 300$

$n =$ _____

14. $54 \div 9 = n$

$n =$ _____

Problem Solving
Exact Numbers or Estimates

Answer each question. Give a reason for your choice.

After more than 25 years of unfriendly relations, two neighboring countries signed a peace agreement. During the 9 days after the signing, people in the two countries celebrated with parades and festivals. About 300,000 families were involved in the celebrations. Everyone was hopeful that the countries would remain friendly forever.

1. Which statement is true?

    a. The countries had been unfriendly for exactly 25 years.

    b. The celebrations lasted for 9 days.

    c. Exactly 300,000 families celebrated.

2. Which word tells you that 300,000 families is an estimate?

    a. after   **b.** hopeful   **c.** about

The 2:00 P.M. movie was almost sold out when Jen arrived at 2:09. Most of the 175 seats in the theater were full. About $1,000 was collected for tickets to that show.

3. Which of the numbers is exact?

    a. The number of people in the theater.

    b. The time Jen arrived.

    c. The amount of money collected.

4. Which might be a good estimate of the number of people at the show?

    a. 100   **b.** 165   **c.** 200

Review and Remember

Solve for *n*.

5. $4 \times n = 16$ **6.** $521 - 23 = n$ **7.** $n + 972 = 12{,}321$

    $n =$ _____ $n =$ _____ $n =$ _____

Using a Calendar

Find each missing number.

1. 21 days = _____ weeks

2. 30 decades = _____ years

3. 6 years = _____ months

4. 2 years = _____ weeks

5. 24 months = _____ years

6. 49 days = _____ weeks

7. 5 decades = _____ years

8. 14 days = _____ weeks

Problem Solving

Use this calendar to answer Problems 9 – 10.

AUGUST						
S	M	T	W	T	F	S
					1	2
3	4	5	6	7	8	9
10	11	12	13	14	15	16
17	18	19	20	21	22	23
24	25	26	27	28	29	30
31						

9. If today is August 4, then what day of the week is 3 weeks and 2 days

from now? _____

10. What will the date be 2 weeks and 4 days after August 12?

_____

Review and Remember

Solve.

11. 783
× 45

12. 32,619
− 845

13. 3,229
+ 123

14. $3\overline{)768}$

Problem Solving
Work Backwards

Work backwards to help you solve the problems.

1. Kyle wants to meet his friends to see a movie at 7:30 P.M.. He has band practice for a half hour beginning at 6:30 P.M. If it takes twenty minutes to travel to the movie theater, does Kyle have enough time? Explain.

_____

_____

2. In June 1992 a sixth-grade class planted a tree in the schoolyard. The tree grew about 3 inches a year. If the tree was 38 inches high in June 1997, about how high was it when it was planted?

_____

3. After the class play ended, Andy sold refreshments for 15 minutes. If he finished at 3:30 P.M. and the play was 1 hour 40 minutes long, what time did the play start?

_____

4. Sean is four months older than Tony. Heather is six months younger than Tony. If Sean's birthday is in April, when are Heather's and Tony's birthdays?

_____

5. A chef is making a dessert that requires one hour to bake, 20 minutes to cool, and 10 minutes to decorate. Then, it must be refrigerated for 90 minutes. If he wants the dessert to be ready at 5:00 P.M., when should he start making it?

_____

Review and Remember

Use paper and pencil, mental math, or a calculator to solve.

6. $20 \times 10 =$ _____   **7.** $624 \div 12 =$ _____   **8.** $56,236 - 1,000 =$ _____

EXPLORE: Measuring With Nonstandard Units

Measure these objects in cubits or spans.

1. length of your foot _____

2. height of your desk _____

3. width of your mathematics textbook _____

4. length of your pencil _____

Problem Solving

5. Amanda's telephone cord is about 3 spans long. Raoul's telephone cord is about 3 cubits long. Whose telephone cord can extend the greatest distance? Why? _____

6. Josiah uses the length of his thumb to measure small objects. About how many thumb lengths long is your hand? _____

_____

Review and Remember

Find each answer. Use paper and pencil or a calculator.

7.	9,898	**8.**	56,639	**9.**	963,942
	845		3,245		21,521
	+ 2,890		+ 36,965		+ 32,412

10.	841	**11.**	102	**12.**	333
	× 13		× 11		× 5

Measuring Length Using Customary Units

Find each length to the nearest $\frac{1}{8}$ inch.

1. |———————————————————| _____ inches

2. |—————————| _____ inches

3. |———————————| _____ inches

Write each missing number.

4. 5 yd = _____ ft

5. 2 mi = _____ ft

6. 3 yd 1 ft = _____ ft

7. 24 ft = _____ yd

Choose the most appropriate unit of measure for the length of each. Write *in.*, *ft*, *yd*, or *mi*.

8. road _____

9. book _____

10. building _____

Problem Solving

11. The distance between John's and Jared's houses is about 5,000 feet. Is that

more than or less than one mile? _____

12. The sign for the Whirlygig ride states that you must be 48 inches tall to ride.

Would a two-year-old child be able to ride? Why or why not? _____

_____

Review and Remember

Find each answer. Use mental math.

13. 350 − 50 = _____

14. 785 + 15 = _____

15. 800 + 200 = _____

16. 50 × 30 = _____

17. 795 + 5 = _____

18. 739 − 609 = _____

19. 999 + 1 = _____

20. 12 × 10 = _____

21. 533 + 167 = _____

Measuring Weight Using Customary Units

Write each missing number.

1. 5 lb = _____ oz

2. 2 T = _____ lb

3. 3 lb 4 oz = _____ oz

4. 6 T = _____ lb

Choose the most appropriate unit of measure for the weight of each.
Write *oz*, *lb*, or *T*.

5. car _____

6. meat _____

7. spoon _____

8. envelope _____

9. cement _____

10. baby _____

Problem Solving

11. An elephant weighs about 3 T. About how many pounds is this?

_____

12. Andria and her family have traveled about 5 miles. Her little sister thinks that
is about 1,000 feet. Andria thinks that is about 26,000 feet. With whom do
you agree? Why? _____

_____

Review and Remember

Round to the underlined place and then estimate each answer.

13. $\underline{3}1 \div 5$

14. $\underline{7}9 - \underline{5}9$

15. $\underline{8}63 + \underline{2}12$

_____

_____

_____

16. $\underline{6}73 - \underline{4}21$

17. $\underline{5}4 \times \underline{3}6$

18. $\underline{8}98 - \underline{1}02$

_____

_____

_____

19. $\underline{4}6 + \underline{9}7$

20. $\underline{4}6 \div 5$

21. $\underline{3}43 \times \underline{1}1$

_____

_____

_____

Measuring Capacity Using Customary Units

Write each missing number.

1. 6 qt = _____ pt

2. 8 qt = _____ gal

3. 48 fl oz = _____ c

4. 6 c = _____ tbsp

Choose the most appropriate unit of capacity. Write *gal, qt, pt, c,* or *tbsp.*

5. bucket _____

6. spoon _____

7. pool _____

8. pitcher _____

9. bathtub _____

10. milk _____

Problem Solving

11. A bathtub fills at the rate of 3 gallons per minute. In 4 minutes, how many quarts will be in the bathtub? _____

12. A recipe requires 2 cups of flour. Kim has a tablespoon, but no measuring cups. What can she do? _____

Review and Remember

Find each answer.

13. 523
   + 56

14. 5,779
   − 632

15. 632
   + 212

16. 43
   × 34

17. 49,356
   − 28,707

18. 512
   × 13

19. 24)24,000

20. 892
   + 14

21. 872 + 18 = _____

22. 543 − 17 = _____

23. 75 × 4 = _____

Measuring Temperature in Degrees Fahrenheit

Find the difference between the two temperatures.

1. 6°F and 60°F _____   **2.** 85°F and 50°F _____

3. 29°F and 66°F _____   **4.** 98°F and 69°F _____

Use each thermometer to find each temperature.

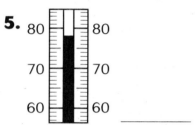   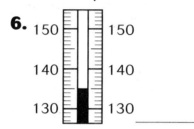   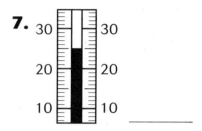

Problem Solving

8. The temperature at noon is 36°F. It was 9°F colder at 8:00 A.M. What was

the temperature at 8:00 A.M.? _____

9. A recipe tells you to start a roast at 500°F, then reduce the heat to 350°F.

By how many degrees do you reduce the heat? _____

10. Water boils at 212°F. Water freezes at 32°F. What is the halfway point

between the boiling temperature and freezing temperature? _____

Review and Remember

Find each answer.

11.   3,621
   +   96

12.   54
   × 6

13.   212
   × 23

14.   789
   − 597

15.   72
   133
  + 15

16.   4,332
  − 998

17.   99
   68
 + 221

18.   201
   × 18

Measuring Length Using Metric Units

Write each missing number.

1. 6 m = _____ cm

2. 8 km = _____ m

3. 3 dm = _____ cm

4. 4 cm = _____ mm

5. 200 cm = _____ m

6. 25 km = _____ m

Choose the most appropriate unit of measure for the length of each. Write *cm*, *dm*, *m*, or *km*.

7. highway _____

8. depth of a pool _____

9. width of a button _____

10. length of a book _____

11. length of a room _____

12. distance to the moon _____

Problem Solving

13. A bulletin board is 1 meter wide. There are 4 drawings to be hung on it that are each 9 centimeters wide. Is there enough room on the bulletin board to hang the drawings side-by-side? Why or why not? _____

_____

_____

14. The pattern for a costume requires 1 meter of a gold trim that costs $0.10 per cm. How much will it cost? _____

Review and Remember

Find each answer.

15. 63 ÷ 9 = _____

16. 9 × 8 = _____

17. 56 ÷ 8 = _____

18. 12 × 8 = _____

19. 88 ÷ 8 = _____

20. 8 × 8 = _____

21. 7 × 6 = _____

22. 9 × 9 = _____

23. 7 × 9 = _____

Measuring Mass Using Metric Units

Write each missing number.

1. 8 kg = _____ g   **2.** 10,000 g = _____ kg

3. 3 kg = _____ g   **4.** 14,000 g = _____ kg

5. 4 kg = _____ g   **6.** 700 g = _____ kg

Choose the most appropriate unit of measure for the mass of each.
Write *g* or *kg*.

7. pencil _____   **8.** car _____

9. toothpaste _____   **10.** furniture _____

11. elephant _____   **12.** raisin _____

Problem Solving

13. The mass of a book is 2,000 g. How many kilograms is that? _____

14. The mass of a bag of flour is 3 kg. How many grams is that? _____

15. The mass of a watermelon is 5 kg. How many grams is that? _____

16. The mass of a small T.V. is 20,000 g. How many kilograms is that? _____

Review and Remember

Find each answer.

17. $45 \times 2 =$ _____   **18.** $122 + 396 =$ _____   **19.** $\$14 \div 2 =$ _____

20. $753 - 164 =$ _____   **21.** $51 \times 3 =$ _____   **22.** $486 + 172 =$ _____

23. $28 \div 4 =$ _____   **24.** $300 - 162 =$ _____   **25.** $19 \times 7 =$ _____

Measuring Capacity Using Metric Units

Write each missing number.

1. 8 L = _____ mL   **2.** 7 L = _____ mL

3. 13 L = _____ mL   **4.** 2,000 mL = _____ L

5. 6 L = _____ mL   **6.** 3,500 mL = _____ L

Choose the most appropriate unit of measure for the capacity of each.
Write *L* or *mL*

7. glass of milk _____   **8.** water in a wading pool _____

9. drops in a dropper _____   **10.** water in a watering can _____

11. liquid soap _____   **12.** shampoo in a bottle _____

Problem Solving

13. A soup recipe requires 1.5 L of water. If you double the recipe, then how many liters of water do you need? _____

14. Juice costs $1.19 per liter. How much change will you get from $5.00 if you buy two liters? _____

15. A fruit punch contains 2L of pineapple juice, 1.5L of orange juice, and 1.5L of sparkling water. How much liquid must the punch bowl be able to hold?

_____

Review and Remember

Find each missing number.

16. $20 + 30 + n = 60$ **17.** $n - 625 = 213$ **18.** $16 \div n = 2$

$n =$ _____ $n =$ _____ $n =$ _____

19. $42 + n + 3 = 50$ **20.** $18 \div 3 = n$ **21.** $814 - n = 6$

$n =$ _____ $n =$ _____ $n =$ _____

Measuring Temperature in Degrees Celsius

Use each thermometer to find the temperature.

1. 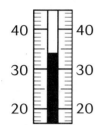   **2.** 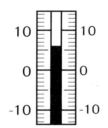   **3.**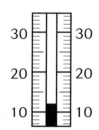

_____   _____   _____

Circle the most appropriate temperature.

4. wearing a swim suit

    a. 14°C   **b.** −3°C   **c.** 34°C

5. cooking soup

    a. 29°C   **b.** 75°C   **c.** 12°C

Problem Solving

6. If it is 24°C at 2:00 P.M. and the temperature was 6°C less at noon, what

was the temperature at noon? _____

7. The temperature increased about 3°C each hour from 8:00 A.M. to
1:00 P.M. At 8:00 A.M. the temperature was 13°C. What is the temperature at

1:00 P.M.? _____

Review and Remember

Find each answer.

8. $20 \times 12 =$ _____   **9.** $6{,}132 + 122 =$ _____   **10.** $25 \times 11 =$ _____

         Use with Grade 4, text pages 284–285.   **83**

Problem Solving
Using Measurement

Solve.

1. The road from Becket to Longmeadow is 132 miles long. Forty-six miles before you reach Longmeadow, you pass through the town of Springfield. A gas station and restaurant are located along the road halfway between Becket and Springfield. How far are the restaurant and gas station from Becket? Draw a picture to explain your answer.

_____

2. A train leaves the station at Alameda twice a day for Broomtown, where it turns around and returns. Each day it travels a total of 280 miles. How many miles apart are Alameda and Broomtown? Explain your reasoning.

_____

_____

3. Yoko's desk is 36 inches wide. She wants to place it along a 5-foot wall with her bookcase, which is 14 inches wide. Will both the desk and bookcase fit along the wall? Draw a picture to explain your reasoning.

_____

_____

Review and Remember

Solve.

4. $15 \div 3 =$ _____   **5.** $48 \div 6 =$ _____   **6.** $8 \times 7 =$ _____   **7.** $9 \times 3 =$ _____

8. $6 \times 6 =$ _____   **9.** $5 \times 8 =$ _____   **10.** $24 \div 8 =$ _____   **11.** $50 \div 5 =$ _____

Mental Math: Dividing Multiples of 10, 100, and 1,000

Divide.

1. 48 ÷ 6 = _____   **2.** 18 ÷ 9 = _____

   480 ÷ 6 = _____   180 ÷ 9 = _____

   4,800 ÷ 6 = _____   1,800 ÷ 9 = _____

Use basic facts and patterns to divide.

3. 300 ÷ 6 = _____   **4.** 120 ÷ 6 = _____   **5.** 450 ÷ 9 = _____

6. 8,100 ÷ 9 = _____   **7.** 4,900 ÷ 7 = _____   **8.** 320 ÷ 4 = _____

9. 900 ÷ 9 = _____   **10.** 20 ÷ 2 = _____   **11.** 60 ÷ 3 = _____

Problem Solving

Use the information in the table to answer Problems 12–13.

Ice Cream Cups	
Flavor	**Number**
Vanilla	720
Chocolate	3,600
Strawberry	90
Butterscotch	180

12. The chocolate flavor will be packaged in packs of 6. How many packages are needed?

_____

13. Strawberry and butterscotch flavors will be combined and packaged together in packs of 3. How many packages are needed? _____

Review and Remember

Solve.

14. 231 × 8 = _____   **15.** 64 ÷ 8 = _____   **16.** 109 − 98 = _____

EXPLORE: Dividing by One-Digit Numbers

Use base-ten blocks to help you divide. Record your work in the chart.

	Number in All	Number of Groups	Number in Each Group	Number Left Over
1.				
2.				
3.				
4.				
5.				
6.				

1. Divide 33 into 4 equal groups.

2. Divide 19 into 2 equal groups.

3. Divide 64 into 8 equal groups.

4. Divide 41 into 8 equal groups.

5. Divide 26 into 5 equal groups.

6. Divide 16 into 5 equal groups.

Problem Solving

Use base-ten blocks to help you solve Problems 7 – 8.

7. Carlos is dividing 7 marbles among 3 friends. How many will each friend get? How many will be left over? _____

8. There are 25 students in Mr. Garcia's fourth-grade class. If five students sit at one table for science, how many tables are needed? _____

Review and Remember

Multiply.

9. $5 \times 2 =$ _____

10. $3 \times 9 =$ _____

11. $8 \times 4 =$ _____

Problem Solving
Interpreting Remainders

Answer each question. Give a reason for your choice.

Teachers at the Washington Street School want to buy new playground equipment. They have $50 to spend on basketballs, which cost $9.00 each. They want to buy 6 basketballs.

1. Which of the following statements is true?

    a. Basketballs cost $50 each.

    b. There is enough money to buy more than two basketballs.

    c. There is not enough money to buy any basketballs.

2. Which of these sentences could you use to find out how many basketballs can be bought?

    a. $50 ÷ $9

    b. $50 × $9

    c. $9 × 6

3. Do the teachers have enough money to buy 6 basketballs? If not, how should you use the remainder from Problem 2 to find out how much more money they need?

    a. Use it to round the quotient up.

    b. Add it to the quotient.

    c. Subtract the remainder from the price of one basketball.

Review and Remember

Find each answer.

4. $4 \times 5 =$ _____   **5.** $7 \times 5 =$ _____   **6.** $4 \times 7 =$ _____   **7.** $9 \times 3 =$ _____

8. $3 \times 8 =$ _____   **9.** $6 \times 7 =$ _____   **10.** $7 \times 7 =$ _____   **11.** $5 \times 6 =$ _____

Dividing Two-Digit Numbers

Divide. Use base-ten blocks to help you.

1. 6)45   **2.** 8)17   **3.** 3)11   **4.** 2)9   **5.** 7)21

6. 4)32   **7.** 5)36   **8.** 9)21   **9.** 5)25   **10.** 8)13

11. 2)13   **12.** 7)14   **13.** 6)43   **14.** 3)22   **15.** 8)79

Problem Solving

Use base-ten blocks to help you solve Problems 16–18.

16. Seventeen fourth-grade students sing in the school chorus. At performances they stand on risers. Each riser holds a group of 9 students. How many groups can stand on one riser? How many students are left over?

_____

17. A juggler has 17 balls to use during his performance. If he divides them into groups of 3 so that the audience can join in the fun, how many groups of

balls does he have? How many are left over? _____

18. Jeff has 13 egg rolls. He divides them equally between himself and 3 of his friends. How many egg rolls does each person get? How many are left over?

_____

Review and Remember

Solve.

19. $111 - 19 =$ _____   **20.** $6 \times 6 =$ _____   **21.** $3,211 + 121 =$ _____

More Dividing Two-Digit Numbers

Divide. Check by multiplying.

1. 2)41 **2.** 6)97 **3.** 3)41 **4.** 8)79 **5.** 7)82

6. 4)62 **7.** 5)75 **8.** 9)89 **9.** 5)63 **10.** 8)85

11. 2)21 **12.** 7)84 **13.** 6)67 **14.** 3)56 **15.** 8)99

Problem Solving

16. Each camping van holds 9 campers.

How many vans are needed for 36 campers? _____

17. A movie pass allows 2 people to see a movie.

How many passes are needed if 32 people want to see a movie? _____

18. Meredith has 13 eggs. She needs 3 eggs for one batch of cookies.

How many batches of cookies can she make? _____

Review and Remember

Find each answer. Use mental math.

19. $100 \div 10 =$ _____ **20.** $630 \div 9 =$ _____ **21.** $140 \div 7 =$ _____

22. $12 \times 10 =$ _____ **23.** $8 \times 30 =$ _____ **24.** $90 \times 40 =$ _____

Estimating Quotients

Estimate each quotient. Tell the numbers you used.

1. 2)15

2. 6)49

3. 3)25

4. 8)57

_____

_____

_____

_____

5. 4)33

6. 5)11

7. 9)460

8. 5)360

_____

_____

_____

_____

9. 6)210

10. 7)230

11. 9)670

12. 3)160

_____

_____

_____

_____

Problem Solving

13. There are 57 party favors in a bag. They are to be divided among 9 party guests. About how many favors will each guest get? _____

14. A piñata has about 73 prizes in it. If 8 people work together to break it, about how many prizes will each person get? _____

15. There are 24 pieces of pizza. Is there enough pizza for 9 people to have 3 pieces each? Explain.

_____

Review and Remember

Find each answer. Use mental math.

16. $762 - 56 =$ _____

17. $845 + 6,321 =$ _____

18. $75 \times 9 =$ _____

Dividing Three-Digit Numbers

Estimate first. Then find the exact answer.

1. 2)152   **2.** 6)348   **3.** 3)426   **4.** 9)558

5. 5)153   **6.** 7)548   **7.** 3)639   **8.** 6)690

9. 4)293   **10.** 2)189   **11.** 3)182   **12.** 9)200

Problem Solving

13. Art supplies cost $189. The cost is to be shared among 9 students.

How much will each student pay? _____

14. Art students share pottery wheels. Seven students are assigned to each pottery wheel. If there are 3 wheels, then how many students are there?

_____

Review and Remember

Find each answer.

15. $821 \times 23 =$ _____   **16.** $3,845 + 6,821 =$ _____

Zeros in the Quotient

Divide. Check by multiplying.

1. 2)402   **2.** 6)648   **3.** 3)326   **4.** 9)958

5. 5)503   **6.** 7)748   **7.** 3)609   **8.** 6)607

9. 4)833   **10.** 2)209   **11.** 3)302   **12.** 9)962

Problem Solving

13. Seth travelled 412 miles in 4 days.

About how many miles did he travel each day? _____

14. A bus carries passengers between two cities twice each day.

At the end of the day, the bus has travelled 432 miles.

How many miles apart are the cities? _____

Review and Remember

Use pencil and paper, mental math, or a calculator to solve.

15. $52,396 \times 89 =$ _____

16. $230 + 30 =$ _____

17. $60 \div 3 =$ _____

18. $7,384 - 2,562 =$ _____

19. $708 + 2,497 =$ _____

20. $250 \div 5 =$ _____

Problem Solving
Write a Number Sentence

Write a number sentence to help you solve each problem. Show your work.

1. In the school talent show there are 3 comedy acts, 8 singing acts, and 4 dance acts. If the talent show lasts one hour, how long can each act perform?

_____

2. Tickets to the talent show cost $0.50. How much will it cost for a family of 5 to attend the show?

_____

3. Refreshments at the talent show cost $0.25 each. If each member of the audience buys an average of one item, and there are 35 people in the audience, how much money is spent on refreshments?

_____

4. Decorations for the talent show cost $15.97. If ticket sales totaled $54.50, how much money was left after paying decoration expenses?

_____

5. One section of the auditorium contains 25 rows of seats. Each row has 28 seats. How many people could be seated in that section?

_____

Review and Remember

Give the value of the underlined digit.

6. 5̲6,792 _____ **7.** 8̲67,324 _____ **8.** 9̲,788 _____

EXPLORE: Averages

Find the average of each set of numbers.

1. 4, 8, 10, 2 ———

2. 3, 21, 3, 9 ———

3. 9, 2, 14, 2, 3 ———

4. 3, 9, 3 ———

5. 11, 16, 9, 13, 1 ———

6. 12, 19, 16, 13 ———

7. 21, 9, 6, 15, 4 ———

8. 1, 1, 14, 6, 3 ———

Problem Solving

Use the table to solve Problems 9–11.

Alisha's Biking Log						
Sunday	Monday	Tuesday	Wednesday	Thursday	Friday	Saturday
12 miles	2 miles	5 miles	5 miles	4 miles	2 miles	12 miles

9. What is the average distance that Alisha biked each day? _____

10. Over which two-day period did Alisha average the greatest number of miles?

The least number of miles? _____

_____

11. On the following Monday, Alisha biked 7 miles. Does this raise or lower her

average? Explain. _____

_____

Review and Remember

Find the value of the underlined digit.

12. 5<u>2</u>,369 _____

13. <u>6</u>2,521 _____

14. 3,<u>6</u>91 _____

Finding Averages

Find the average of each set of numbers.

1. 2, 3, 6, 1 _____

2. 6, 5, 2, 3, 4 _____

3. 16, 19, 29, 63, 23 _____

4. 15, 5, 36, 25, 14 _____

5. 102, 65, 22, 103 _____

6. 32, 84, 76, 104, 4 _____

7. Find the average cost for each item. Complete the chart.

Cost of Each Item	Total Cost	Number of Items	Average Cost per Item
$2, $4, $6			
10¢, 14¢ 25¢, 7¢			
$25, $16, $24, $15			
7¢, 5¢, 2¢, 5¢, 6¢			

Problem Solving

8. Juan's scores on a board game are 23, 26, 8, 20, 12, and 7.

What is his average score? _____

9. The Piper Fife and Drum Club is raising money for uniforms.
So far, they have contributions of $3, $5, $15, $2, and $5.

What is the average contribution? _____

Review and Remember

Solve.

10. 6)306

11. 48 ÷ 6 = _____

12. 5,603 − 209 = _____

Problem Solving
Using Data

Use the data in the pictograph to solve the problems.

Books in the School Library	
Category	**Number of Books**
Fiction	▯ ▯ ▯ ▯ ▯ ▯ ▯
Science	▯ ▯ ▯ ▯ ▯ ▯ ▯ ▯ ▯
History	▯ ▯ ▯ ▯ ▯ ▯ ▯
Art	▯ ▯ ▯ ▯ ▯
Music	▯ ▯ ▯ ▯ ▯ ▯ ▯ ▯ ▯ ▯ ▯ ▯ ▯

Each ▯ stands for 3 books.

1. What is the average number of books in each category? _____

2. Which type of book represents the mode? _____

3. If 21 new art books are added to the collection, will there be a change in the mode? Why or why not?

_____

_____

4. If each fiction book is worth about $2.00, about how much is the fiction

collection worth? _____

Review and Remember

Solve.

5. 78,897 + 4,005 = _____

6. 897,878 − 3,476 = _____

7. 14 × 9 = _____

8. 42.63 ÷ 7 = _____

Parts of a Region

Write the fraction for the shaded part of each region.

1.

_____

2.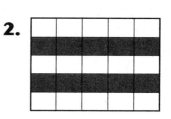

_____

Draw a picture to show each fraction.

3. $\frac{1}{3}$

4. $\frac{9}{10}$

5. $\frac{5}{5}$

6. $\frac{3}{4}$

7. $\frac{5}{8}$

8. $\frac{3}{6}$

Problem Solving

9. Amy and Jonathan are making a round spinner for a game. The spinner is to be divided into four equal parts. Sketch the spinner.

10. Jamaica designs square floor tiles. Each square is divided in half. Sketch two possible designs she could use.

Review and Remember

Use mental math to solve.

11. 20 + 50 = _____

12. 400 + 200 = _____

13. 800 − 300 = _____

Parts of a Set

Write the fraction that represents the following parts of the
seven-piece set of shapes shown here.

1. Triangles _____

2. Squares _____

3. Circles _____

4. Circles and squares _____

Write a number to complete each sentence.

5. To find $\frac{1}{3}$ of 21, divide 21 into _____ equal parts.

6. To find $\frac{1}{5}$ of 15, divide 15 into _____ equal parts.

Find the fractional part of each number.

7. $\frac{1}{3}$ of 12 _____

8. $\frac{1}{4}$ of 8 _____

9. $\frac{1}{6}$ of 18 _____

10. $\frac{1}{7}$ of 14 _____

Problem Solving

11. One-half of the students in the fourth grade are visiting the Earth exhibit at
the science museum, and one-half are visiting the weather exhibit. There are
20 fourth-grade students. How many will visit each exhibit?

_____

12. If there are 21 beads on an abacus, and $\frac{1}{3}$ of the beads are brown, how

many beads are brown? _____

Review and Remember

Find each answer.

13. $3,045 + 396 =$ _____

14. $13 \times 121 =$ _____

15. $5,402 - 776 =$ _____

16. $3,200 \div 80 =$ _____

EXPLORE: Equivalent Fractions

Use fraction pieces to decide if the fractions are equivalent. Write *yes* or *no*.

1. $\frac{2}{3}, \frac{4}{5}$ _____

2. $\frac{3}{6}, \frac{1}{2}$ _____

3. $\frac{1}{4}, \frac{2}{5}$ _____

4. $\frac{2}{3}, \frac{4}{6}$ _____

Use your fraction pieces to write at least one equivalent fraction for each fraction.

5. $\frac{1}{3}$ _____

6. $\frac{1}{4}$ _____

7. $\frac{2}{5}$ _____

8. $\frac{3}{5}$ _____

Problem Solving

9. In a pattern, $\frac{2}{3}$ of the numbers are whole numbers. If there are 12 numbers in the pattern, how many of them are whole numbers? _____

10. A recipe calls for $\frac{1}{2}$ cup of blueberries. Seth's measuring cup is divided into fourths. How many fourths should Seth measure? _____

11. Marty has $\frac{1}{2}$ of a dozen eggs. Marcia has $\frac{6}{12}$ of a dozen eggs. Who has more eggs? _____

Review and Remember

Write the value of the underlined digit.

12. 5̲6,789 _____

13. 7̲21,930 _____

14. 62,3̲69 _____

15. 6,72̲1 _____

16. 34,8̲92 _____

17. 999,99̲9 _____

Equivalent Fractions

Tell whether each fraction is written in simplest form. Write *yes* or *no*. Then use fraction strips to write an equivalent fraction.

1. $\frac{2}{4}$ _____

2. $\frac{1}{3}$ _____

3. $\frac{4}{5}$ _____

4. $\frac{3}{12}$ _____

5. $\frac{2}{8}$ _____

6. $\frac{3}{9}$ _____

7. $\frac{1}{5}$ _____

8. $\frac{3}{8}$ _____

9. $\frac{2}{6}$ _____

Write each fraction in simplest form. Use fraction strips if needed.

10. $\frac{2}{10}$ _____

11. $\frac{3}{6}$ _____

12. $\frac{4}{6}$ _____

13. $\frac{3}{9}$ _____

14. $\frac{4}{12}$ _____

15. $\frac{3}{8}$ _____

Problem Solving

16. Kerri thinks that eating three slices of a pizza that is cut into sixths is the same as eating $\frac{1}{2}$ pizza. Brian thinks that eating three slices is the same as eating $\frac{1}{3}$ pizza. Who do you think is right? Explain your reasoning.

_____

17. How should Daphne divide 12 inches of ribbon if she wants six equal

pieces? four equal pieces? _____

_____

Review and Remember

Find each answer.

18. $72 \times 9 =$ _____

19. $6,790 - 209 =$ _____

20. $640 \div 8 =$ _____

21. $810 \div 90 =$ _____

22. $5,432 + 135 =$ _____

23. $21 \times 32 =$ _____

Problem Solving
Reasonable Answers

Answer each question. Circle the letter of the correct answer.

David is directing a play to raise money for the food bank. He has spent $50 to use the school auditorium, $23 to make copies of his script, and $67 on materials for the set. Props cost $19. On opening night 45 people saw the play and on closing night 52 people saw the play. Tickets cost $5 per person.

1. Which number sentence tells how many people saw the play?

   a. 50 + 23 + 67 + 19

   b. 52 − 45

   c. 45 + 52

2. Is $60 a reasonable estimate of David's expenses?

   a. Yes; expenses, when rounded, total $60.

   b. No; expenses, when rounded, are more than $100.

   c. No; expenses, when rounded, are less than $20.

3. Which statement describes how much money David gave to the food bank?

   a. Ticket sales minus expenses, or $326.

   b. Number of people plus expenses, or $256.

   c. Expenses times number of performances, or $118.

Review and Remember

Give the value of the underlined digit.

4. 6,684 _____   **5.** 632,521 _____

6. 56,123 _____   **7.** 6,785 _____

Comparing and Ordering Fractions

Compare. Write $>$, $<$, or $=$. Use fraction pieces or draw pictures to help you.

1. $\frac{2}{3}\bigcirc\frac{4}{6}$ **2.** $\frac{1}{4}\bigcirc\frac{2}{3}$ **3.** $\frac{4}{5}\bigcirc\frac{1}{3}$ **4.** $\frac{2}{6}\bigcirc\frac{1}{3}$

5. $\frac{1}{2}\bigcirc\frac{3}{10}$ **6.** $\frac{1}{9}\bigcirc\frac{1}{3}$ **7.** $\frac{1}{4}\bigcirc\frac{2}{3}$ **8.** $\frac{1}{2}\bigcirc\frac{7}{10}$

Order each set from least to greatest. Use fraction pieces to help you.

9. $\frac{1}{2}, \frac{1}{3}, \frac{1}{4}$ _____   **10.** $\frac{1}{6}, \frac{2}{3}, \frac{1}{8}$ _____

11. $\frac{3}{5}, \frac{1}{3}, \frac{1}{2}$ _____   **12.** $\frac{3}{4}, \frac{7}{8}, \frac{1}{2}$ _____

Problem Solving

The graph shows materials for sale by the museum recycling center. Use the graph for Problems 13–14.

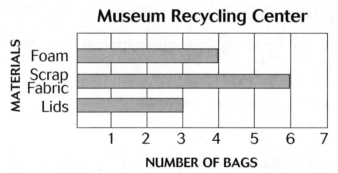

Museum Recycling Center

13. Foam pieces cost $0.29 per bag. How much will the foam pieces cost?

_____

14. A teacher buys all of the lids and scrap fabric for a class project. Lids cost $0.39 per bag and scrap fabric costs $0.19 per bag. How much will the

teacher pay altogether? _____

Review and Remember

Find each missing number.

15. $15 + 13 + n = 75$ **16.** $13 \times n = 39$ **17.** $900 - n = 300$

$n =$ _____ $n =$ _____ $n =$ _____

Mixed Numbers

Draw a picture to find the missing numbers.

1. $\frac{3}{2} =$ _____ $\frac{1}{2}$

2. $\frac{7}{5} =$ _____ $\frac{2}{5}$

3. $\frac{11}{4} =$ _____ $\frac{3}{4}$

4. $\frac{16}{3} =$ _____ $\frac{1}{3}$

5. $\frac{9}{2} =$ _____ $\frac{1}{2}$

6. $\frac{9}{4} =$ _____ $\frac{1}{4}$

Write an improper fraction and a mixed number for each picture.

7.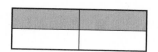

_____

8.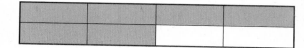

_____

Problem Solving

9. Laura cut 18 triangles for a quilt. Each block requires 4 triangles. How many blocks can she make? How many triangles will be left over?

_____

10. Two fourth-grade classes are touring an art museum. There are 35 students to be divided into four groups. How many students will be in each group? How many students will be left over?

_____

Review and Remember

Find each quotient. Check by multiplying.

11. $324 \div 3 =$ _____

12. $320 \div 5 =$ _____

13. $\$12 \div 4 =$ _____

EXPLORE: Add and Subtract Fractions with Like Denominators

Use fraction pieces to add or subtract. Write each answer as a mixed number or whole number when possible.

1. $1\frac{1}{4} + \frac{1}{4} =$ _____

2. $\frac{3}{8} + 2\frac{1}{8} =$ _____

3. $\frac{5}{6} + \frac{1}{6} =$ _____

4. $\frac{3}{4} - \frac{1}{4} =$ _____

5. $\frac{5}{6} - \frac{1}{6} =$ _____

6. $\frac{2}{3} - \frac{1}{3} =$ _____

7. $\frac{7}{10} + \frac{9}{10} =$ _____

8. $\frac{9}{10} - \frac{1}{10} =$ _____

Draw a picture to show each number sentence.

9. $\frac{1}{2} + \frac{1}{2} + \frac{1}{2} + \frac{1}{2} = 2$

10. $\frac{5}{6} - \frac{1}{6} = \frac{4}{6}$ or $\frac{2}{3}$

Problem Solving

11. Sam biked $\frac{1}{2}$ mile on Monday, one mile on Tuesday, and $\frac{1}{2}$ mile on Wednesday. How many miles did he bike altogether? _____

12. A bookcase requires shelves that are $\frac{3}{4}$ foot long. How many feet of shelving are required for two shelves? _____

Review and Remember

Follow the rule to complete each table.

	Rule: Multiply by 4	
	Input	Output
13.	512	
14.	223	
15.	615	

	Rule: Divide by 6	
	Input	Output
16.	636	
17.	954	
18.	732	

Addition of Fractions and Mixed Numbers

Add. Write each sum in simplest form. Use fraction pieces to help you.

1. $\frac{1}{6} + \frac{5}{6} =$ _____

2. $\frac{3}{8} + \frac{1}{8} =$ _____

3. $2\frac{2}{3} + 1\frac{1}{3} =$ _____

4. $\frac{4}{11} + \frac{2}{11} =$ _____

5. $3\frac{1}{12} + \frac{4}{12} =$ _____

6. $4\frac{9}{13} + \frac{12}{13} =$ _____

7. $5\frac{3}{8} + 2\frac{1}{8} =$ _____

8. $4\frac{1}{4} + 7\frac{2}{4} =$ _____

9. $7\frac{1}{8} + 1\frac{1}{8} =$ _____

10. $2\frac{1}{7} + 2\frac{2}{7} =$ _____

11. $6\frac{1}{4} + 1\frac{3}{4} =$ _____

12. $6\frac{1}{6} + 2\frac{3}{6} =$ _____

Use mental math. Find each sum.

13. $\frac{1}{6} + \frac{1}{6} =$ _____

14. $\frac{3}{8} + \frac{2}{8} =$ _____

15. $\frac{7}{12} + \frac{1}{12} =$ _____

Problem Solving

16. You studied $\frac{1}{2}$ hour during school, $\frac{1}{2}$ hour on the bus, and $1\frac{1}{2}$ hours at home. How long did you study altogether? _____

17. Draw a picture that shows why $\frac{1}{3}$ of a cake is greater than $\frac{1}{4}$ of a cake.

Review and Remember

Use mental math to find each answer.

18. $400 + 200 =$ _____

19. $2,500 \div 50 =$ _____

20. $1,000 - 200 =$ _____

Name _____

Subtraction of Fractions and Mixed Numbers

Subtract. Write each answer in simplest form. Use fraction pieces, if you like.

1. $\frac{5}{6} - \frac{1}{6} =$ _____

2. $\frac{5}{8} - \frac{2}{8} =$ _____

3. $2\frac{2}{3} - 1\frac{1}{3} =$ _____

4. $1\frac{10}{11} - \frac{2}{11} =$ _____

5. $3\frac{5}{12} - 1\frac{4}{12} =$ _____

6. $4\frac{2}{3} - 1\frac{2}{3} =$ _____

7. $5\frac{3}{8} - 2\frac{1}{8} =$ _____

8. $4\frac{1}{4} - 2\frac{1}{4} =$ _____

9. $7\frac{5}{6} - 1\frac{1}{6} =$ _____

10. $4\frac{3}{7} - 2\frac{2}{7} =$ _____

11. $6\frac{3}{4} - 2\frac{1}{4} =$ _____

12. $6\frac{4}{9} - 2\frac{3}{9} =$ _____

Use mental math. Find each difference.

13. $\frac{3}{6} - \frac{1}{6} =$ _____

14. $3\frac{7}{8} - 1\frac{1}{8} =$ _____

15. $8\frac{2}{3} - 4\frac{1}{3} =$ _____

16. $1\frac{3}{4} - \frac{1}{4} =$ _____

Problem Solving

17. A farmer has $2\frac{2}{3}$ acres. If he plants corn in $\frac{1}{3}$ acre, how much is left for other

crops? _____

18. A poster is $1\frac{1}{3}$ yards wide. If it is placed on a wall that is $5\frac{2}{3}$ yards wide, then

how much space will be left for another poster? _____

19. Samantha has $5\frac{3}{4}$ yards of fabric. If she uses $2\frac{1}{2}$ yards for curtains and $1\frac{1}{4}$

yards for a pillow case, how much fabric does she have left? _____

Review and Remember

Estimate each answer.

20. $87 \times 3 =$ _____

21. $395 \div 4 =$ _____

22. $600 - 291 =$ _____

Problem Solving
Guess and Check

Guess and check to solve these problems.

1. One cup of juice and 4 corn dogs cost $4.50. Corn dogs cost twice as much as juice. How much does each item cost?

_____

2. Two student desks and one teacher's desk are against a wall. The width of 2 student desks equals the width of one teacher's desk. The width of all 3 desks measures 100 inches. How wide is each type of desk?

_____

3. Will is 5 years older than Brian. Amanda is 10 years older than Will. The ages of the 3 people add up to 50 years. How old is each person?

_____

4. For every blue square Taryn uses in her pattern, she uses 3 yellow rectangles and 2 green triangles. She has 18 shapes in the pattern so far. How many blue, yellow, and green pieces does she have? Draw a picture of a pattern that explains your answer.

_____

Review and Remember

Find the elapsed time.

5. 8:15 A.M. to 12:15 P.M.

_____

6. 9:30 P.M. to 6:30 A.M.

_____

7. 5:00 A.M. to 1:00 P.M.

_____

8. 3:00 P.M. to 10:00 P.M.

_____

EXPLORE: Adding Fractions with Unlike Denominators

Use fraction pieces or draw pictures to find each sum.

1. $\frac{1}{6} + \frac{2}{3} =$ _____

2. $\frac{1}{8} + \frac{1}{4} =$ _____

3. $\frac{1}{5} + \frac{3}{10} =$ _____

4. $\frac{1}{2} + \frac{1}{3} =$ _____

5. $\frac{7}{12} + \frac{1}{3} =$ _____

6. $\frac{1}{3} + \frac{1}{12} =$ _____

7. $\frac{3}{8} + \frac{1}{2} =$ _____

8. $\frac{1}{3} + \frac{1}{9} =$ _____

9. $\frac{1}{4} + \frac{1}{2} =$ _____

Draw pictures to explain each number sentence.

10. $\frac{1}{2} + \frac{1}{3} = \frac{5}{6}$

11. $\frac{1}{3} + \frac{1}{6} = \frac{3}{6}$ or $\frac{1}{2}$

12. $\frac{1}{8} + \frac{1}{4} = \frac{3}{8}$

Problem Solving

13. A recipe calls for $\frac{1}{4}$ cup lima beans and $\frac{2}{3}$ cup pinto beans. How many

cups of beans are there altogether? _____

14. To trim a costume, you need $\frac{1}{2}$ yard at the neck and $\frac{1}{6}$ yard at the wrist.

How much trim is needed? _____

Review and Remember

Use a pencil and paper, mental math, or a calculator to solve.

15. $90 \times 5 =$ _____

16. $1,081 \div 47 =$ _____

17. $412 - 199 =$ _____

18. $13 \times 15 =$ _____

19. $5,436 + 389 =$ _____

20. $54,000 \div 90 =$ _____

EXPLORE: Subtracting Fractions with Unlike Denominators

Use fraction pieces or draw pictures to find each difference.

1. $\frac{1}{2} - \frac{1}{3} =$ _____

2. $\frac{1}{3} - \frac{1}{4} =$ _____

3. $\frac{1}{5} - \frac{1}{10} =$ _____

4. $\frac{2}{3} - \frac{1}{6} =$ _____

5. $\frac{1}{2} - \frac{3}{8} =$ _____

6. $\frac{5}{8} - \frac{1}{2} =$ _____

7. $\frac{3}{4} - \frac{1}{2} =$ _____

8. $\frac{2}{3} - \frac{1}{9} =$ _____

9. $\frac{3}{4} - \frac{1}{12} =$ _____

Draw pictures to explain each number sentence.

10. $\frac{1}{3} - \frac{1}{6} = \frac{1}{6}$

11. $\frac{1}{4} - \frac{1}{8} = \frac{1}{8}$

12. $\frac{1}{2} - \frac{1}{8} = \frac{3}{8}$

Problem Solving

13. A chili recipe calls for $\frac{3}{4}$ tablespoon chili powder. For a less spicy chili, a person should use about $\frac{1}{8}$ tablespoon less. About how much chili powder is required for a less spicy chili? _____

14. Miguel walked $\frac{7}{10}$ mile to Jake's house. This is about $\frac{1}{5}$ mile less than the distance to Erika's house. How far is Erika's house from Miguel's house?

_____

Review and Remember

Compare. Write $>$, $<$, or $=$.

15. $\frac{1}{2} \bigcirc \frac{1}{4}$

16. $\frac{1}{4} \bigcirc \frac{1}{3}$

17. $\frac{3}{6} \bigcirc \frac{1}{2}$

18. $\frac{5}{12} \bigcirc \frac{1}{3}$

Problem Solving
Using Circle Graphs

Use the information in the circle graph to solve the problems.

Cost of Erien's Art Supplies

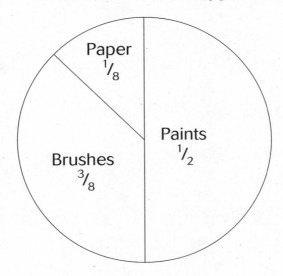

1. Write a number sentence to show the cost of the three types of art supplies

   that Erien bought. _____

2. Which item was the most expensive? _____

3. Which item was the least expensive? _____

4. What fraction of Erien's expenses are made up of paints and paper? _____

Review and Remember

Solve for *n*.

5. $6 \times n = 54$ **6.** $n \div 9 = 9$ **7.** $5 \times 9 = n$ **8.** $51 \div n = 17$

$n =$ _____ $n =$ _____ $n =$ _____ $n =$ _____

EXPLORE: Fractions and Decimals

Draw a model that shows each fraction or decimal.

1. $\frac{4}{10}$

2. 2.30

3. $1\frac{6}{10}$

4. $\frac{9}{10}$

5. 1.5

6. $4\frac{2}{10}$

Problem Solving

7. Sally planted $\frac{2}{10}$ of her flower garden with yellow flowers, $\frac{5}{10}$ with pink flowers, $\frac{1}{10}$ with purple flowers, and $\frac{2}{10}$ with red flowers. Draw a model and color it to show what her garden looked like. Then label each part with a decimal. Which section is largest? _____

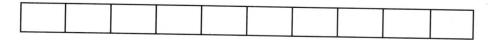

8. Ko made clay beads for a necklace. $\frac{3}{10}$ of the beads have circle designs, $\frac{2}{10}$ are plain, and $\frac{5}{10}$ have line designs. Draw a model to show what the beads looked like.

Review and Remember

Use mental math to solve.

9. $150 + 50 =$ _____

10. $1,600 + 400 + 200 =$ _____

Name _____

Relating Mixed Numbers and Decimals

Write each as a decimal.

1. $\frac{3}{10}$ _____

2. $\frac{2}{100}$ _____

3. $6\frac{6}{10}$ _____

4. $2\frac{9}{10}$ _____

5. $5\frac{8}{10}$ _____

6. $8\frac{7}{100}$ _____

7. $9\frac{27}{100}$ _____

8. $7\frac{21}{100}$ _____

9. $8\frac{71}{100}$ _____

Write each decimal as a fraction.

10. 0.3 _____

11. 1.07 _____

12. 5.19 _____

13. 8.98 _____

14. 11.4 _____

15. 0.05 _____

Problem Solving

16. Edwina bought two boxes of pencils. Each box had 10 pencils. She kept one box of pencils for herself, and gave away 2 of the pencils in the other box. Write a mixed number and a decimal to tell how many boxes of pencils

Edwina had left. _____

17. Simon bought a package of 100 sheets of construction paper. He gave 14 sheets to his sister and 23 sheets to a friend. Write a fraction and a decimal

to tell what part of the package he gave away. _____

18. Maria is writing a 10-page paper. She has written 7 pages. She says she has written 0.8 of the paper. Her mother says she has written $\frac{7}{10}$ of the paper.

Who is correct? _____

_____

Review and Remember

Find the answers.

19. $852 \times 23 =$ _____

20. $1,526 - 896 =$ _____

21. $9\overline{)74}$

Decimal Place Value

Write each as a decimal.

1. five and six tenths _____   **2.** eleven and twelve hundredths _____

3. seven and eight tenths _____   **4.** six and nine tenths _____

Write the word form for each decimal.

5. 32.98 _____

6. 669.4 _____

7. 821.32 _____

Tell the place-value of each underlined digit.

8. 34.2<u>5</u> _____   **9.** 456.8<u>7</u> _____   **10.** 5<u>1</u>2.97 _____

Problem Solving

11. A decimal number has 8 in the tens place, 4 in the ones place, 6 in the hundredths place, and 1 in the tenths place. Write the number. _____

12. Used skates cost $13.97. How much money is shown in the hundredths place? _____

13. I am a number. The digit in my hundredths place is double the digit in my tenths place. The digit in my tenths place is 1 less than the digit in my ones place. The digit in my ones place is 3. What number am I? _____

Review and Remember

Solve by using paper and pencil, mental math, or a calculator.

14. $89 \times 52 =$ _____   **15.** $520 - 20 =$ _____

16. $84 \div 4 =$ _____   **17.** $563 - 109 =$ _____

Comparing and Ordering Decimals

Compare. Write $>$, $<$, or $=$.

1. 2.3 $\bigcirc$ 3.1

2. 7.16 $\bigcirc$ 7.1

3. 32.4 $\bigcirc$ 32.40

4. 32.12 $\bigcirc$ 31.12

5. 65.9 $\bigcirc$ 65.09

6. 8.32 $\bigcirc$ 8.3

7. 9.39 $\bigcirc$ 9.32

8. 62.29 $\bigcirc$ 62.92

9. 10.1 $\bigcirc$ 10.10

Order each set of numbers from greatest to least.

10. 32.12, 32.21, 32.11 _____

11. 2.59, 2.95, 2.90 _____

12. 0.1, 0.001, 0.0001, 1.0 _____

13. 4.32, 3.86, 4.02, 3.99 _____

Problem Solving

Use the table to solve Problems 14–15.

Speed in Miles per Hour		
Vehicle	**7:00 A.M.**	**7:00 P.M.**
Bus	37.2	45.19
Car	37.91	45.23

14. Which vehicle had the greatest speed at 7:00 P.M.? _____

15. When did a vehicle travel slower than 45.3 miles per hour?

_____

Review and Remember

Find each answer.

16. $7\overline{)336}$

17. $\frac{4}{9} + \frac{3}{9} =$ _____

18. $\frac{7}{11} - \frac{3}{11}$ _____

Problem Solving
Multistep-Problems

Answer each question. Give a reason for your choice.

Craig's father has a car that he bought in 1996. Before that it belonged to Craig's uncle for 2 years. It was used for 3 years before that by Craig's grandmother. The car was bought by Craig's grandfather 4 years before that.

1. What could you do to find the year that Craig's grandmother began using the car?

   a. Subtract 2 years from 1996.

   b. Subtract 4 years from 1994.

   c. Subtract 5 years from 1996.

2. How could the answer to Problem 1 help you know when Craig's grandfather bought the car?

   a. Subtract 4 years from the answer.

   b. Subtract 9 years from the answer.

   c. Add 4 years to the answer.

3. When did Craig's grandfather buy the car? _____

Review and Remember

Compare. Write $<$, $>$, or $=$.

4. $6 + 19 \bigcirc 4 + 12$ **5.** $7 \times 3 \bigcirc 4 \times 5$ **6.** $45 - 7 \bigcirc 30 + 8$

7. $12 \div 3 \bigcirc 4 + 1$ **8.** $3 \times 8 \bigcirc 4 \times 6$ **9.** $8 + 17 \bigcirc 17 - 8$

10. $68 - 3 \bigcirc 50 + 10$ **11.** $63 \div 9 \bigcirc 8 \times 0$ **12.** $81 \times 1 \bigcirc 9 \times 9$

Using Rounding to Estimate

Round each decimal to the nearest whole number.

1. 22.3 _____   **2.** 63.21 _____   **3.** 56.95 _____   **4.** 84.62 _____

Write two decimals that round to each number.

5. 75 _____   **6.** 31 _____   **7.** 84 _____   **8.** 21 _____

Round to the nearest whole number. Then add or subtract.

9.　231.12　　**10.**　404.02　　**11.**　861.42　　**12.**　523.62
　　 + 841.96　　　　 − 186.13　　　　 + 562.12　　　　 −　23.69

Write the value in decimals. Then round to the nearest dollar.

	Dollars	Quarters	Dimes	Pennies	Decimals	Rounded
13.	2	3	0	0		
14.	5	0	4	2		
15.	8	3	5	0		

Problem Solving

16. It is 75.2 miles from Lincoln to Las Santos, and an additional 2.9 miles from Las Santos to Birket. About how far is it from Lincoln to Birket?

_____

17. Amy is buying ingredients for a salad. Lettuce costs $1.49; celery costs $1.19; carrots cost $1.79; and dressing costs $2.97. About how many

dollars will the ingredients cost? _____

Review and Remember

Find each equivalent measure.

18. 5 yd = _____ ft   **19.** 3 ft = _____ in.   **20.** 2 m = _____ cm

Problem Solving
Draw a Picture

Make a drawing to solve Problems 1–3.

1. In a line for tickets, Keith is ahead of Amy. Misha is behind Amy. Corinne is ahead of Keith. How are they lined up, from front to back?

_____

2. One wall has 5 lockers. Those belonging to Gus and Sam are on the ends. Matt's is next to Gus' and Phil's is next to Sam's. Where is Philana's locker?

_____

3. Stuart is building a dog pen. Every 5 feet, he must place a support post. If he has 60 feet of fencing to make a square pen, where will he place the support posts? How many posts should he buy?

_____

Review and Remember

Round to the nearest whole number.

4. 23.4 _____

5. 56.89 _____

6. 73.24 _____

Adding Decimals

Estimate first. Write your estimate to the right of each problem.

Then find each sum.

1.　0.2
　　+ 0.6 ____

2.　0.89
　　+ 0.36 ____

3.　2.3
　　+ 4.9 ____

4.　6.05
　　+ 5.21 ____

Find the sums.

5.　56.23
　　+ 5.45

6.　0.56
　　+ 2.36

7.　2.89
　　+ 0.36

8.　4.60
　　+ 97.36

9.　52.63
　　+ 8.36

10.　2.23
　　+ 11.23

11.　74.19
　　+ 0.84

12.　3.65
　　+ 57.20

Problem Solving

Use the chart to answer Problems 13–14.

13. What was the total rainfall for the spring and summer of 1957?

_____

14. About how much more rain in the spring and summer did El Camino have in 1997 than in 1977?

_____

Rainfall in El Camino			
Season	**1957**	**1977**	**1997**
Spring	0.34 in.	1.23 in.	1.87 in.
Summer	4.74 in.	2.12 in.	4.36 in.

Review and Remember

Find each answer.

15. $2{,}121 \div 3 =$ _____

16. $\frac{1}{6} + \frac{1}{2} =$ _____

17. $790 \times 4 =$ _____

Subtracting Decimals

Estimate first. Then find each difference.

1. 0.9
  − 0.4 _____

2. 1.72
  − 0.16 _____

3. 8.3
  − 4.9 _____

4. 6.05
  − 3.11 _____

Find each difference.

5. 54.13
  − 5.45

6. 3.56
  − 0.26

7. 7.60
  − 0.26

8. 4.60
  − 1.32

9. 2.83
  − 1.36

10. 24.23
  − 3.69

11. 89.14
  − 0.85

12. 6.35
  − 0.57

Problem Solving

13. Thirty and four-tenths inches of yarn is needed for a lion puppet. If 12.2 inches of that is for the tail, how much is needed for the rest of the puppet?

_____

14. Various vegetables weighing 3.4 pounds, 1.2 pounds, 6.4 pounds, and 11.4 pounds are purchased. A bag holds 25 pounds of vegetables. How many more pounds of vegetables can be added to the bag?

_____

Review and Remember

Find each answer.

15. $6 \times 87 =$ _____

16. $7\overline{)96}$

17. $\frac{4}{7} - \frac{1}{7} =$ _____

18. $1.2 + 4.5 =$ _____

19. $382 - 94 =$ _____

20. $11 \times 200 =$ _____

Name _____

Adding and Subtracting Decimals

Estimate first. Then find each sum or difference.

1. 5.4
− 1.4 ____

2. 4.32
+ 0.16 ____

3. 9.3
− 2.9 ____

4. 11.05
+ 3.21 ____

Find each sum or difference.

5. 54.1
5.45
+ 0.26

6. 3.56
2.5
+ 10.26

7. 37.6
25.6
+ 51.32

8. 4.60
6.32
+ 56.10

9. 42.83 − 6.35 = _____

10. 1.23 − 0.95 = _____

11. 165.3 − 65.23 = _____

12. 849.2 − 95.32 = _____

Problem Solving

13. A turkey farmer has 989.72 pounds of turkey at the beginning of November. At the end of the month, he has 12.9 pounds left. How many pounds did the farmer sell?

_____

14. At the beginning of the turkey season, each turkey weighed an average of 9.2 pounds. By the middle of the season, each turkey weighed an average of 18.96 pounds. What was the average weight gain of each turkey?

_____

Review and Remember

Find each equivalent measure.

15. 20 yd = _____ ft

16. 4 ft = _____ in.

17. 300 cm = _____ m

Problem Solving
Comparing Prices

Compare prices to solve Problems 1–4.

1. Cartons of yogurt cost $0.69 each. A package containing 6 cartons costs $3.00. Which is the best buy? Why? _____

_____

2. Ms. Sanchez has a coupon for 40 cents off the price of a dozen oranges. A dozen oranges cost $3.49. Apples are on sale for $2.99 a dozen. Which fruit is the best buy? Why? _____

_____

3. Ace Film charges $0.25 per print to develop film. Bargain Outlet charges $4.80 to develop 24 prints. Which is the better buy for a roll of 24 prints? Why? _____

_____

_____

4. An 8-pack of juice costs $3.84. A 6-pack of a different brand costs $3.12. Which is the better buy? Why?

_____

_____

Review and Remember

Add or subtract.

5.   45.2
   13.24
$+ 34.2$

6.   53.32
   4.31
$+ 23.63$

7.   17.31
$- 4.30$

8.   14.92
$- 3.36$

EXPLORE: Polygons

Which figures below are polygons? Write yes or no.

1.

_____

2.

_____

3.

_____

Which figure does not belong? Explain your reasoning.

4. a.

b.

c.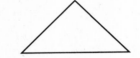

_____

Problem Solving

5. Carrie is arranging sticks to make shapes. She has made a closed figure,

with 10 sides. How many vertices are there? _____

6. Niki's younger brother thinks that because a circle is a closed shape, it is a polygon. What would you tell his brother?

_____

_____

Review and Remember

Use mental math to solve.

7. $40 \times 60 =$ _____   **8.** $1{,}200 \div 10 =$ _____   **9.** $20 \times 30 =$ _____

Points, Lines, Segments, Rays, and Angles

Draw an example of each.

1. line *DE*

2. line segment *RS*

3. ray *AB*

4. angle *ABC*

5. two parallel lines

6. angle with vertex *B*

Tell whether each angle is right, obtuse, or acute.

7.

_____

8.

_____

9.

Problem Solving

10. What kind of angles do the hands of a clock make at 3:00 and 9:00?

_____

11. Beth thinks that train tracks run parallel to each other. Kim thinks that they run perpendicular to each other. With whom do you agree? Why?

_____

_____

Review and Remember

Give the value of the underlined digit.

12. 45.2̲4 _____

13. 7̲83.15 _____

Problem Solving
Spatial Reasoning

Answer each question. Give a reason for your choice.

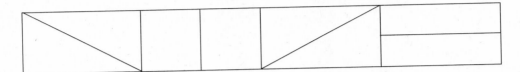

1. How many triangles are there in the picture? _____

2. How many right angles are there in the picture? _____

3. Draw the next 4 repeats of the pattern.

4. Which group of shapes can be combined to make a square? _____

a.

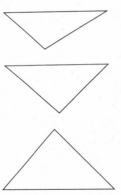

b.

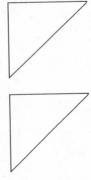

Review and Remember

Identify each angle.

5.

6.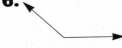

7.

_____ _____ _____

Circles

Use the circle to complete Problems 1– 3.

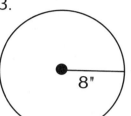

1. radius = _____

2. diameter = _____

3. The circumference is about _____.

8"

Draw the given radius or diameter in each circle.
Then estimate each circumference.

4.

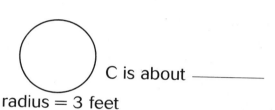

C is about _____

radius = 3 feet

5.

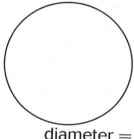

C is about _____

diameter = 10 feet

Problem Solving

6. A Ferris wheel has a diameter of 30 feet. What is its radius?

What is its approximate circumference? _____

7. A merry-go-round has a circumference of approximately 100 feet.
If horses are positioned approximately 3 feet apart, then about

how many horses are there? _____

Review and Remember

Find each answer.

8. $\frac{1}{2} + \frac{1}{3} =$ _____

9. $1.25 + 3.58 =$ _____

10. $5.36 - 2.69 =$ _____

11. $\frac{3}{4} - \frac{3}{5} =$ _____

Name _____

Congruent Figures

Are the figures in each pair congruent? Write yes or no. Explain.

1.

_____

2.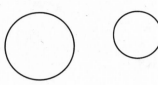

_____

3.

_____

4.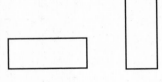

_____

5. Draw two congruent figures.

6. Draw two figures that are not congruent.

Problem Solving

7. Three puzzle pieces are triangles. One has an obtuse angle, one has a right angle, and one has 3 acute angles. Are the pieces congruent? Why or why not?

_____

_____

8. Suppose you are asked to design a floor with tiles having congruent shapes. Sketch a design that you might use.

Review and Remember

Estimate each answer, then solve.

9. $2,604 - 789 =$ _____

10. $965 + 369 =$ _____

Similar Figures

Draw a similar figure for each.

1.

2.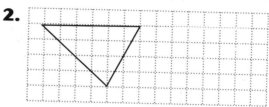

3.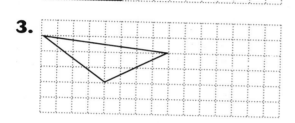

4.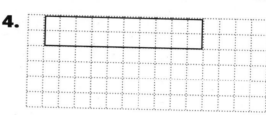

Problem Solving

5. You are making a scale drawing of a garden plot. Will the scale drawing and the actual garden plot be similar figures? Why or why not?

_____

6. How would you explain to a young student the difference between congruent and similar figures? Make drawings to explain your ideas.

_____

_____

_____

Review and Remember

Use a calculator to find *n*.

7. $3{,}458 + n = 24{,}980$

$n =$ _____

8. $632 \times 56 = n$

$n =$ _____

9. $n \div 59 = 75$

$n =$ _____

EXPLORE: Symmetry

Draw lines of symmetry if you can.

1.

2.

3.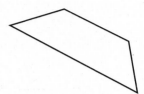

4. Draw a figure with a line of symmetry.

5. Draw a figure that does not have a line of symmetry.

Problem Solving

6. Heather is making a nature symmetry matching game. She wants to use drawings of butterflies, snowflakes, and pine cones.

Which of these can she use as symmetrical figures? _____

7. Yuneng is folding paper to make an origami figure. If he folds a square piece of paper in half, and then in half again, how many symmetrical figures has he made? Make a drawing that explains your answer. _____

Review and Remember

Find each answer.

8.   9.26
     + 236.3

9.   92.369
     − 26.236

10.   36.236
     523.3
     + 56.32

EXPLORE: Slides, Flips, and Turns

Tell how each figure was moved. Write *slide*, *flip*, or *turn*.

1.   _____

2. 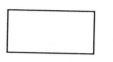 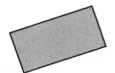 _____

3. 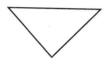  _____

Problem Solving

4. Andy is making a pinwheel design for a quilt. He wants each part of the pinwheel to be congruent triangles. How can he use flips, slides, or turns

to help him? _____

_____

5. Print your first name inside a box. Then make a flip, slide, or turn of it.

Review and Remember

Find each answer.

6. $725 \times 2 =$ _____

7. $892 \times 51 =$ _____

8. $895 \div 5 =$ _____

9. $1{,}518 \div 6 =$ _____

Problem Solving
Use Logical Reasoning

Use logical reasoning to solve each problem.

1. What number am I? _____

- My ten-thousands digit is 4 less than my thousands digit.
- My thousands digit is 3 times my hundreds digit.
- My hundreds digit is $\frac{1}{2}$ my tens digit.
- My tens digit is 6.
- My ones digit is 3 more than my tens digit.

2. What are the two mystery numbers? _____

- The numbers are prime.
- All digits are odd.
- When the two digits of each are added together, they equal 16.
- The numbers are less than 100.
- They are greater than 78.

3. What is the mystery number sentence? _____

- Its factors are divisible by 3.
- Its factors are divisible by 10.
- Its product is greater than 2,500.
- Its product is less than 3,000.

Review and Remember

4. Draw a flip, slide and turn of this figure.

Perimeter

Find the perimeter of each figure.

1.

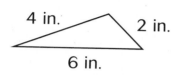

4 in.   2 in.

6 in.

_____

2.

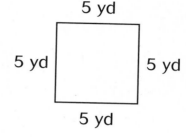

5 yd

5 yd   5 yd

5 yd

_____

3.

16 ft

2 ft   2 ft

16 ft

_____

4.

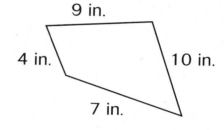

9 in.

4 in.   10 in.

7 in.

_____

Problem Solving

5. If each side of a stop sign is 10 inches, then what is the perimeter of a stop sign? Make a labeled drawing to explain your answer. (**Hint**: a stop sign is an octagon.)

_____

6. The perimeter of a YIELD sign is 60 inches. Two sides are each 24 inches.

How long is the third side? _____

YIELD

Review and Remember

Use mental math to find each quotient.

7. 500 ÷ 5 = _____

8. 300 ÷ 10 = _____

9. 200 ÷ 50 = _____

Area

Find each area in square units.

1.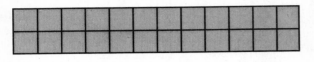

_____

2.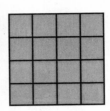

_____

3. Write the initial of your first name in block letters on the grid. How many square units is your initial? _____

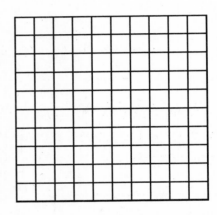

Problem Solving

4. Megan wants to tile a floor that is 16 feet long and 8 feet wide.

If she uses 1-foot square tiles, how many tiles will she need? _____

5. Megan decides to use tiles that are 2 feet long and 2 feet wide.

How many tiles will she need to tile the floor? _____

Review and Remember

Find each answer.

6. $800 - 632 =$ _____

7. $72 \div 9 =$ _____

8. $\$7.35 + \$8.98 =$ _____

9. $3{,}112 \times 20 =$ _____

10. $63 \div 9 =$ _____

11. $\$52.30 - \$14.95 =$ _____

Problem Solving
Using Perimeter and Area

Use perimeter and area to solve each problem.

Community Gardens

15 yd

45 yd

1. The town is going to divide the Community Gardens into rectangular plots that are 3 yd × 15 yd. How many plots will the Community Gardens have? Draw the plots in the figure above to help you. _____

2. What is the area of each plot? _____

3. The town charges the gardener of each plot $1.24 per yard to fence the entire garden. How much does the town charge the gardeners altogether? _____

4. How many yards of fencing will a gardener with an end plot have to pay for?

_____

Review and Remember

Use mental math to find each answer.

5. 1,200 ÷ 30 _____   **6.** 60,000 ÷ 2,000 _____

7. 100,000 ÷ 500 _____   **8.** 90,000 ÷ 3,000 _____

9. 4,500 ÷ 150 _____   **10.** 36,000 ÷ 1,200 _____

Space Figures

Name and draw an object that looks like each shape.

1. cone

2. cube

3. cylinder

4. sphere

Label each figure.

5. 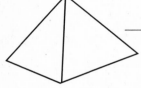 _____

6.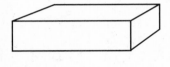

_____

Problem Solving

7. Which of the following shapes can roll? Which can not roll?
cylinder, rectangular prism, sphere, square pyramid

_____

_____

8. During Math Olympics Day, the fourth-grade class is having a space-figure
scavenger hunt. Which of the following should they not include on their list?
Why? cube, cone, circle, square, square pyramid, rectangle, rectangular prism

_____

Review and Remember

Compare. Use $>$, $<$, or $=$.

9. $81 \div 9 \bigcirc 100 \div 10$

10. $270 \div 30 \bigcirc 240 \div 80$

EXPLORE: Volume

Multiply to find the volume of each figure. Use cubes to check your answer.

1.

_____

2.

_____

3.

_____

4.

_____

Problem Solving

5. Habib is helping organize his brother's blocks. He has assembled one set into 3 rows, each 3 rows high and 4 rows deep. He organized another set into 5 rows, each 2 rows high and 5 rows deep. How many blocks has he

organized?_____

6. Jonathan has a solid object with a volume of 36 cubic units. Draw a possible object and give its dimensions. How many figures can you make?

_____

_____

Review and Remember

Solve.

7. $4 \times 9 =$ _____   **8.** $8 \times 7 =$ _____   **9.** $4 \times 8 =$ _____   **10.** $7 \times 7 =$ _____

Mental Math: Dividing by Multiples of Ten

Use basic facts to divide. Write the basic fact and find each quotient.

1. 80 ÷ 20

2. 400 ÷ 40

3. 140 ÷ 70

_____

_____

_____

4. 50 ÷ 10

5. 90 ÷ 30

6. 180 ÷ 90

_____

_____

_____

Find each quotient. Check by multiplying.

7. 60)360

8. 40)160

9. 30)120

10. 20)160

11. 40)280

12. 50)250

13. 70)350

14. 80)720

Problem Solving

15. Each family that goes to the Happy Times Amusement Park on family day pays $30. After one hour, a ticket booth has collected $6,000. How many

families have passed through the ticket booth in one hour? _____

16. A ferry traveling between the mainland and an island carries 1,800 passengers each day from noon to 6:00 P.M. About how many

passengers does it carry each hour? _____

17. A hotel can hold about 240 people. There are 80 rooms in the hotel. About

how many people can each room hold? _____

Review and Remember

Use mental math or a calculator.

18. 50 × 50 = _____

19. 1,545 ÷ 15 = _____

20. 80 × 30 = _____

Estimating Quotients

Use compatible numbers to estimate each quotient. Tell which compatible numbers you used.

1. $820 \div 22$

_____

2. $360 \div 92$

_____

3. $148 \div 21$

_____

4. $352 \div 74$

_____

5. $921 \div 33$

_____

6. $728 \div 92$

_____

7. $29 \overline{)180}$

8. $41 \overline{)243}$

9. $37 \overline{)124}$

10. $52 \overline{)365}$

_____

11. $49 \overline{)153}$

12. $57 \overline{)369}$

13. $72 \overline{)565}$

14. $77 \overline{)639}$

_____

Problem Solving

15. There are 8,190 picture books on 91 shelves in the children's library. If each shelf holds about the same number of books, about how many books are on

each shelf? _____

16. The children's room at the library wants to buy enough shelving to hold 5,606 books. Each shelf will hold about 70 books. About how many shelves

are needed? _____

Review and Remember

Find each answer.

17. $34 \times 125 =$ _____   **18.** $563 + 36 + 852 =$ _____   **19.** $506 - 37 =$ _____

Problem Solving
Choose the Operation

Answer each question. Give a reason for your choice.

The Tate Elementary School held its Spring Concert on Friday. Forty-five students participated. Thirty of the students were members of the chorus. The remaining students played instruments including the piano, recorder, flute, and violin.

1. How many students played instruments?

_____

2. If 2 students played piano, 3 students played flute, and the remaining students were evenly divided between recorder and violin, how many students played each of these instruments?

_____

3. In the chorus, half the singers were altos. Ten were sopranos. The others were soloists. How many soloists were there?

_____

4. On the stage, the entire chorus was arranged on three rows of risers. How many singers were on each riser?

_____

Review and Remember

Use a calculator to find each answer. Round each quotient to the nearest tenth.

5. $45 \overline{)456,213}$ **6.** $231 \overline{)23,785}$ **7.** $87 \overline{)56,214}$ **8.** $347 \overline{)234,712}$

9. $29 \overline{)1,288}$ **10.** $362 \overline{)44,825}$ **11.** $71 \overline{)93,128}$ **12.** $201 \overline{)65,884}$

Dividing by Two-Digit Numbers: One-Digit Quotients

Divide. Check by multiplying.

1. $61\overline{)549}$ **2.** $52\overline{)257}$ **3.** $37\overline{)149}$ **4.** $72\overline{)652}$

5. $23\overline{)163}$ **6.** $46\overline{)242}$ **7.** $72\overline{)436}$ **8.** $92\overline{)851}$

Problem Solving

9. There are 24 servings in a box of 144 crackers. How many crackers are in each serving?

_____

10. A box can hold 96 items that weigh 8 ounces each. How much will a box full of these items weigh? (Hint: 16 ounces is equivalent to one pound.)

_____

11. There are 235 people marching in the parade. If there are 47 rows, how

many people are in each row? _____

Review and Remember

Solve for n.

12. $762 - n = 263$ **13.** $n \times 17 = 85$ **14.** $1,593 - n = 269$

$n =$ _____ $n =$ _____ $n =$ _____

EXPLORE: Adjusting Your Estimate

Estimate each quotient. Then divide and adjust your quotient if necessary.

1. $72\overline{)649}$ _____

2. $54\overline{)328}$ _____

3. $38\overline{)238}$ _____

4. $81\overline{)169}$ _____

5. $47\overline{)402}$ _____

6. $64\overline{)329}$ _____

7. $21\overline{)159}$ _____

8. $93\overline{)396}$ _____

Problem Solving

9. A carton holds 24 glasses. A restaurant purchased 200 glasses. How many

cartons are needed to hold the glasses? _____

10. A hotel needs 177 towels, which are only sold in packages of one dozen. If each package costs $10.00, then about how much will the hotel pay for the towels? Explain your reasoning.

_____

_____

11. Andrea made $218 one month mowing lawns. She mowed 17 lawns. She

says she made about $15 per lawn. Is she correct? Explain. _____

_____

Review and Remember

Find the sum or difference.

12. $\dfrac{3}{4} - \dfrac{1}{4} =$ _____

13. $\dfrac{5}{6} - \dfrac{4}{6} =$ _____

14. $3\dfrac{3}{4} + \dfrac{1}{4} =$ _____

Problem Solving
Choose a Strategy

Use a strategy to solve each problem.

Make a Graph	Find a Pattern
Guess and Check	Write a Number Sentence
Draw a Picture	Make a List

1. The distance around a corn field is about one mile. Gate posts are positioned every 1,000 feet. How many posts are needed?

_____

2. The art teacher is rearranging the tables in his room. Instead of combining 2 square tables to make rectangles that seat 12 students, he is leaving them as separate square tables. How many students can sit at each square table?

_____

3. Grades 4, 5, and 6 attended Harmony Day in the city. Each student could choose 2 different souvenirs from a collection of hats, pennants, mugs, and pencils. How many combinations are there?

_____

Review and Remember

Find each answer.

4. $50 × 3 _____

5. 4)$\overline{\$16}$

6. $2.31 × 4 _____

7. $8.69 + $0.34 _____

Name _____

Dividing by Two-Digit Numbers: Two-Digit Quotients

Divide. Check each by multiplying.

1. $72\overline{)867}$ **2.** $65\overline{)911}$ **3.** $37\overline{)600}$ **4.** $47\overline{)614}$

5. $58\overline{)648}$ **6.** $81\overline{)976}$ **7.** $29\overline{)913}$ **8.** $12\overline{)155}$

Problem Solving

9. Jen earned $425 mowing lawns. If she charges $25 per job, how many jobs did she have? _____

10. Travis has taken karate lessons for 30 months. His certificate states that he has had 150 lessons. How many lessons did he average each month? _____

11. There were 546 people at the play. Each row in the theater has 12 seats. How many rows were completely filled? _____

Review and Remember

Find each answer.

12. $82 \times 14 =$ _____   **13.** $2,463 + 5,120 =$ _____

14. $8,420 - 5,142 =$ _____   **15.** $980 \div 7 =$ _____

16. $1,333 + 666 =$ _____   **17.** $17 \times 36 =$ _____

Problem Solving
Using Operations

Solve.

1. Jake's family drove 529 miles on their vacation. Their van gets about 23 miles per gallon of fuel. The average cost of fuel was $1.31. About how much did the family pay for fuel on their vacation?

_____

2. Ron is having his bike repaired. Materials cost $23.98. Labor costs $12.00 per hour. If the repairs take one hour and 15 minutes, then how much does Ron owe altogether?

_____

3. A jacket is on sale for $23.87. This is $5.12 less than the original price. What was the original price?

_____

4. A music store is having a special sale. Every customer who buys 3 or more items receives a free carrying case. If 277 carrying cases were given away during the special sale, then what is the least number of items the store sold?

_____

5. Select apples cost $3.60 a dozen. Premium apples cost $0.35 each. Which is the better buy? Explain your reasoning.

_____

_____

Review and Remember

Find each answer.

6. $\frac{1}{2} + \frac{1}{2}$ _____

7. $\frac{1}{3} + \frac{1}{3}$ _____

8. $\frac{1}{5} + \frac{2}{5}$ _____